I0072827

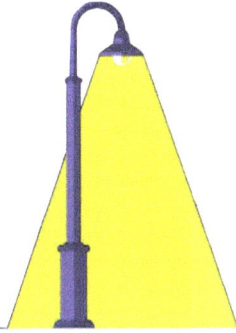

RATIONAL MYOPIA

HOW CAPITAL MARKETS LEARN

KENT OSBAND

Rational Myopia: How Capital Markets Learn
Copyright © 2020 by Kent Osband. All Rights Reserved.

No part of this publication may be reproduced, stored in a retrieval system or transmitted, in any form or by any means—electronic, mechanical, photocopying, recording or otherwise—without prior written permission from the publisher, except for the inclusion of brief quotations in a review.

For information about this title or to order other books and/or electronic media, contact the publisher:

Kent Osband
Mountain Brook, AL
www.risktick.com
info@risktick.com

Hardcover ISBN: 978-1-7343376-0-0
Paperback ISBN: 978-1-7343376-1-7
Electronic ISBN: 978-1-7343376-2-4

Printed in the United States of America
Cover design: 1106 Design

First published 12 March 2020
Last updated 24 December 2020

Contents

Part III: Credit Markets — 111

Part IV: Equity Markets — 163

Preface

Capital markets make big bets on things we can't see based on projections of risks we don't know. Latecomers to human history, they have grown over the past two centuries into prime movers of global wealth. They are revered for affirmations of wealth and reviled for bubbles and crashes. They are lauded for promoting decentralized investment and lambasted for poor judgment.

Capital markets seem chronically inconsistent. Prices are generally more volatile than the income streams they discount but the excess is highly uneven. Sometimes long-term valuations look unduly complacent. Sometimes long-term valuations seem overwhelmed by fear and despair. Future returns are heavily discounted even though competition favors maximization of long-term wealth.

Remarkably, all these manifestations flow from variants of a single equation, which describes rational learning under uncertainty. The Core Learning equation doesn't presume knowledge of future trends like orthodox finance does. It doesn't presume endemic irrationality like behavioral finance does. Instead, it treats market participants as rational gamblers, who infer changing future trends from the limited information at hand and bet on them with hopes of long-term enrichment.

From this perspective, capital markets are neither visionary prophets nor ships of fools. They are decentralized, mostly rational social learners who can't sample the future risks they aim to predict. As a result, they're torn between focusing on noisy current news that might be unreliable and tapping older evidence that might have ceased to be relevant. I call that tension rational myopia.

Rational myopia is the banner of a new paradigm in finance. I counterpose it both to the rational expectations emphasized by orthodox theorists and the irrational exuberance emphasized by behaviorists. On the contrary, it helps unify them by demonstrating that rational learning can take both calm and turbulent forms.

Unfortunately, the Core Learning equation isn't easy to grasp. The simple formula is a shorthand for a potentially infinite set of updates every instant, linked in a chain reaction that can inflate small differences quickly into big ones and let big differences dissipate. A similar difficulty arose in the analysis of fluid dynamics. The driving Navier-Stokes equation was known for a century before physicists were convinced that it explained physical turbulence. What persuaded them were supercomputer simulations of Navier-Stokes that looked unmistakably turbulent.

Inspired by this experience, I focus in this book on simulations. If a picture is worth a thousand words, most of this book is valued in pictures. I also try to minimize the math. Readers can get through most of the first half on basic algebra; the equations serve mainly as formal definitions of terms used in the text. In the second half, the equations get more complex and I can't talk around them without causing more confusion than I avoid. To compensate, I fit the parameters to historical data, which lets readers learn a lot just by looking at the charts and gauging their resemblance to observed market behavior.

My main qualifications for writing this book are the window seats I've had on major financial crises and the perspective I bring to analyzing them. The crises matter because they reveal some market innards that are normally hidden from view. The perspective comes from trying to meld theory and practice in areas where theorists and practitioners have long been at odds.

I was once a radical critic of capital markets and the capitalist system they epitomize. I stopped out of Harvard, swung a sledgehammer for two years at an Alabama steel mill and tried to organize for socialist revolution. Eventually I earned a PhD in economics from UC Berkeley, learned Russian, and studied in what was then the Soviet bloc. My enlightenment was very disillusioning. As rough as competitive markets can be, command economies are worse. So wasteful, so stultifying, so prone to corruption. Only suppression of dissent and brute state power can keep them from falling apart. I stumbled into finance by accident, when an investment bank I had never heard of sought a specialist on the newly ex-Soviet economies.

Joining Goldman Sachs in 1994, I knew so little about what drove capital markets that I didn't know not to say I didn't know. Searching for insight, I browsed the finance aisles of bookstores serving Wall Street. While a few shelves contained books from respected academics, many more shelves touted so-called technical analysis of past price history, aka charting. I found this shocking. Bookstores near pharmaceutical labs don't sell alchemy recipes. Bookstores near NASA don't sell incantations for magic carpets. Why were skilled Wall Street traders lured by something that most academics ridicule? A related puzzle was the huge emphasis on networking. What did traders hope to learn from each other that they couldn't find out better on their own?

Personally, I've never been a good networker or chart reader. I'm more interested in figuring out what's truly right or wrong than in gathering others' opinions or divining market moods. For short-term forecasting, this is more vice than virtue. Most capital market prices are far too complex for a single mind to grasp. One needs to network with others and learn to weight their claims appropriately. While that runs the risk of getting caught in an echo chamber, even the trend in the echo chamber can provide useful information. The most valuable quality any market forecaster can offer is a knack for being slightly ahead of the curve.

At the time I didn't see any of this clearly. However, I did experience a moment of enlightenment in December 1994 when the Mexican peso devalued. My better-connected colleagues had been adamant it would not. Emerging debt markets tanked worldwide on fears of contagion. For the first time I realized that I knew something the market consensus didn't, namely that there was no need to treat Mexican debt as a benchmark. Within a few months Polish spreads broke through the imagined Mexican ceiling and never looked back. I moved on to head strategy for emerging markets debt trading at Credit Suisse First Boston. Later I became chief portfolio strategist for a global macro hedge fund at Fortress Investment

Group. Over and over I was struck by the incongruities between the studiousness of traders, the speed of market adjustments to short-term news, and the market's challenge in seeing the longer-term big picture.

So much for the practical roots of my tangles with rational myopia. The theoretical roots stretch back to my graduate studies, where I wrote my PhD dissertation on incentives for good forecasts. Joining Wall Street shifted my attention to crisis risks in capital markets. Conventional mean-variance models tend to fail badly in crisis. To mitigate this, it is useful to distinguish between risks that are shared by a broad class of assets and risks that behave like random noise. The shared risks are often highly fat-tailed and can't be adequately summarized by means and variances. I wrote my first book Iceberg Risk to explain this.

The answers posed a bigger puzzle: why are aggregate capital market risks so highly fat-tailed? Exploration led me to Mordecai Kurz's work on rational beliefs and Robert Liptser and Albert Shirayev's work on optimal filtering of random processes. For several years I wrote a regular column on these topics for Wilmott magazine. The efforts helped me learn what I was trying to explain and revealed the cumulant hierarchy that drives rational turbulence. After the financial crash in 2008/9, the post-mortem criticisms struck me as focusing far too much on lender errors and not enough on building robustness to errors. To help counter that, I wrote a second book Pandora's Risk emphasizing that uncertainty in finance is bound to induce big tail risks.

In hindsight, I underestimated central banks' resolve to offset the collapse of a mortgage-led bubble by financing a bigger sovereign debt bubble. In general, the ability of capital markets to find reassurance is at least as remarkable as their ability to take fright. The common denominator is the myopic focus on recent data. I myself didn't appreciate how myopic until I ran simulations. I started writing about this five years ago, but early drafts bogged down in math and minutia. Eventually I let the simulations take center stage and show their colors. This book is the result. By ironic coincidence, it goes to press just as markets are crashing with high turbulence on news of a spreading coronavirus epidemic.

Abbreviations

AI	artificial intelligence		F	market E
APL	Almost-Perfect Learner		f	Kelly fraction
B	bond price		G,g	percent / log growth rate
BP	Bayesian Purist			
BR	Beta Reset		GARCH	Generalized Autoregressive Conditional Heteroskedasticity
bps	basis points			
C,c	credit spreads or consumption		GDP	gross domestic product
CMA	cumulative moving average		h,h^*	objective / stationary probability density
CP	Classical Purist		iid	independent, identically distributed
CRRA	constant relative risk aversion		ITM	inter-threaded market
D_{eff}	effective defaults or gamma shape		J	pure dividend yield
			K,k	capital / share of total capital
E	mean belief			
ELH	efficient learning hypothesis		log	logarithm / logarithmic
EMH	efficient market hypothesis		M,N	pool size
			NPV	net present value
EMA	exponential moving average		OECD	Organisation of Economic Cooperation and Development

p	subjective probability
P	ratings transition matrix or equity price
PL	Perfect Learner
Q,q	price-to-dividend ratio / multiplier
R,r	percentage/log discount rate
RMSE	root mean squared error
S	survival rate or Sharpe ratio
s	diffusion speed
std	standard deviation
S&P	Standard & Poor's
TD	Tiny Doubters
TD2L	Tiny Doubter 2-layer
TD3L	Tiny Doubter 3-layer
T,t	time
u_i	exit rate from grade i
T_{eff}	effective sample size
U	information measure
US	United States
V	variance of beliefs
vol	volatility
x	observed outcome
y	standardized Brownian noise
z	other evidence or Brownian noise

Greek Characters

α,β	beta distribution parameters
γ	relative risk aversion
Δ	change
η	learning rate
θ	Bernoulli risk
I	identity matrix
κ	cumulant
K	cumulant generating function
λ	perceived reset rate
Λ	actual reset rate
μ	drift of Brownian x
M	moment generating function
ν	aggregate default risk
ξ	consumption/savings tradeoff multiplier
π	reset distribution or Markov switching rate
Π	switching rate matrix
ρ	rate of time preference
σ	volatility of x
ς	equity price volatility
ϕ,ϕ_L	evidence/logistic function
ψ	intemporal elasticity of substitution
ω	credit risk multiplier

PART I:
BRIDGING DIVIDES

Capital markets are neither knowledge machines nor ships of fools. They are social learners, continually correcting and refining their flawed predictions. Since no one can sample the future, agents study the historical record, look for relevant trends and project their continuation. When the evidence confirms a stable trend, learning is smooth and strengthens consensus. When the evidence dramatically changes, rational learners are torn between dismissing it as outlier and embracing it as a new norm. Rapid learning breeds rational turbulence.

Markets in short-term securities tend to perform like optimal learners, with hypotheses weighted by capital rather than credibility. Yet that doesn't make them good long-term forecasters. Corrective evidence about long-term trends arrives slowly and tomorrow's payoffs depend far more on changes in others' beliefs than on changes in fundamental trends.

The combination induces rational myopia, which works akin to natural selection in biology. Stability spurs dominance by a narrow selection of views. Turmoil favors adaptability and variation.

Wisdom versus Folly

Capital refers to ownership claims on future proceeds. There are basically two ways to assess its worth. One method estimates the expected prices of all future revenues and expenses, discounts them for delay and risk, and aggregates them into net present value (NPV). The other method lets capital markets trade claims. Ideally, the NPV "fair price" should coincide with the "market price" equating supply and demand. In practice, there is massive noise and it is hard to probe for meaning. How well do capital markets capture the proverbial wisdom of crowds? How badly do they capitulate to the proverbial madness of crowds?

Rival findings have split academic finance into two schools. The orthodox school emphasizes investor rationality, reinforced by market pressures that separate fools from their money. It treats markets as knowledge machines, which accurately foresee future risks and respond optimally to them. The behaviorist school emphasizes human frailties, the huge swings in market valuations compared to fundamentals, and skittish responses to news. It treats markets as ships of fools, careening between complacency and panic.

In natural science, rival principles battle until one school triumphs or a higher view bridges the divide. For example, is light a particle or a wave? The two interpretations clashed for centuries, as there was ample evidence to support both interpretations and no apparent way to reconcile them. Eventually quantum theory managed to explain wave/particle duality, but only by. changing our notion of what "is" is: photons and other subatomic particles are only probabilistically here or there.

In contrast, orthodox and behaviorist schools of finance have settled into an uneasy truce. The 2013 Nobel Prize in economics was jointly awarded to behaviorist Robert Shiller, orthodox Eugene Fama and the mostly orthodox Lars Hansen. The message is ecumenical toleration. If rationality explains a puzzling phenomenon, invoke it. If not, blame a human foible. That way, every empirical finding supports one of the schools without refuting the other. Apparently, *homo economicus* is schizoid, with reasonable and foolish sides that capital markets mediate between.

Most finance theorists applaud the truce since each school garners acclaim and there's less pressure to conform to either. Most practitioners like it too, since it encourages pragmatism. However, the truce rests on conceptual quicksand. From an evolutionary perspective, competitive market pressures ought to drive foolishness to the margins. Even if most investors drink from an everlasting fountain of wrongheadedness, why can't some develop immunity or learn to avoid it?

Another strange aspect is the restricted scope of wrongheadedness. In many practical applications, orthodox models positing full rationality work well. Yet aggregate valuations often seem way off the mark, e.g., price-to-earnings ratios for equities that oscillate to extremes far greater than subsequent earnings histories justify. Why aren't long-term forecasts calmer and more reasonable than short-term forecasts?

Here is another physics analogy, which comes closer to capturing the challenge facing finance. Look at Figure 1.1. A submarine is slicing through calm seas and leaving behind a turbulent wake. How many kinds of water does it reveal? If we count calm black water as one, raging white water as two, and perhaps the thin intermediate layer of grey water as three, the total is … one. One kind of liquid water, with two main kinds of flow: smooth and turbulent.

Figure 1.1: Turbulent wake of a submarine[1]

Turbulent white water doesn't just look different from smooth black water. It feels different to anyone stuck in it. Its unpredictability unnerves. From a naïve perspective, something enters the water and stirs it up. Most ancient cultures believed in moody gods of the seas. The best known is Poseidon, who made sailing the Aegean Sea a gamble on tranquility or torment. Not until the nineteenth century did physicists discover the true cause of turbulence, and even they didn't understand why it behaved as it did. It took another hundred years to convince the profession, thanks to supercomputer simulations, that a single equation could generate behavior that looked realistic (Ecke 2005).

Too bad modern finance wasn't there to sort things out. Looking at analogues of Figure 1.1, orthodox theorists would focus on black water. They would emphasize how smoothly the submarine slices through it and how quickly the wake subsides. Behaviorists would focus on white water. They would emphasize how precarious calm is and how much commotion the submarine stirs. Figure 1.2 summarizes the what-if analysis.

Figure 1.2: If finance viewed water the way it views homo economicus

Irrationally Exuberant Water

Rational Calm Water

I apologize for the hyperbole. In truth, this book wouldn't be possible without the insights of orthodox and behavioral economists. I was just venting frustration at the dualism. Like in physics, a single equation can explain both kinds of flows. I call it the Core Learning equation.

I didn't discover this equation. Walter Wonham (1964) did, although only its simplest form. Robert Liptser and Albert Shiryaev (1977) discovered the more general form and proved its optimality for a broad range of problems. Nor am I the first to apply this equation to finance. Alexander David (1997) used it to price equities that switch between two possible rates of dividend growth. A decade later Lubos Pastor and Pietro Veronesi (2009) surveyed a host of articles focused on learning in finance, and many of them applied the equation too.

Still, the Core Learning equation hasn't garnered nearly the attention it deserves. That's because even people who know what it is think it applies only to convergence of opinions in calm black water. They don't see the connection to discord and divergence in turbulent white water. That connection is key to bridging the orthodox/behaviorist divide.

Core Themes

There is no doubt that some people foresee some future trends well and manage to parlay that into riches. There is also no doubt that people are chronically irrational. Still, to bridge the divide I won't presume either. Instead I draw on three related propositions:

P1:   Core trends occasionally change without clear identification.

P2:   Market participants try to rationally infer future trends from past and current outcomes.

P3:   Competition prods toward maximization of long-term wealth.

P1 acknowledges that core drivers of profit are potentially unstable and cannot be deduced from symmetries or other first principles. P2 affirms traders' dedication to learning from experience. P3 measures evolutionary fitness by accumulated capital and gambling prowess. As we shall see, P3 helps select for rationality in P2, P2 enhances the impact of P1, and P1 stirs turmoil in P2 and P3.

From a learning perspective, capital markets are dominated by rational, weakly-informed gamblers. The gamblers are uncertain about current trends because evidence is noisy and because trends occasionally change in ways that make past evidence irrelevant. They are even less certain about future trends because no one can sample the future. Like drunks who search under lampposts for their car keys although they lost them somewhere else, capital markets obsess over current news because that's as close to the future as they can see. Evidence tends to get processed efficiently for short-term securities. However, risk aversion and deference to consensus slow or distort the repricing of long-term securities. Trading on price momentum helps counteract slow adjustment but can drive long-term valuations to extremes.

In short, capital markets are rationally myopic: great short-term error correctors and poor long-term predictors. That's puzzling, since rationality begs for prudent foresight. But how should we approach a future shrouded in darkness? Shouldn't we inspect current evidence even more closely for hints of old trends subsiding and new trends emerging? Shouldn't we network more closely with others, who might see something we don't? That's what rationally myopic people do.

Imagine you're driving a car backward and all you see clearly is the road already traveled. Wouldn't you look more closely at recent twists and turns, to gauge whether you're heading off course? Wouldn't you watch other cars more carefully and incline to imitate their moves? Wouldn't you be more tentative and make frequent corrections, many of which you quickly reverse? That's what market prediction is like.

Orthodox finance counters that the twists and turns don't matter, that they're random noise around a stable trend. To maintain our bearings, we just need to map the trend and follow it. The guiding equations and parameters are analogous to those for industrial materials under stress, for which reference handbooks are readily available. Why aren't analogous handbooks available for finance?

It might be argued that financial engineering coefficients are so precious that major banks and hedge funds keep the information private. Yet turnover in finance is high and prized secrets encourage snooping. Why hasn't some disgruntled insider or hacker revealed the true values and why do so many finance quants spend so much time re-estimating them? The only plausible answer is that the values aren't stable.

This helps explain why market traders spend so much time studying price charts. Why obsess over what people used to believe about the future and how that has changed? However, data on economic fundamentals looks backward too. The further we want to forecast into the future, the less relevant past fundamentals become. If price trends are more predictable and have greater impact on near-term changes, it makes sense to bet on those. Similarly, networking to discern others' beliefs may reveal more profit opportunities than studying long-term growth projections.

Here are six more puzzling features of capital markets. From an orthodox perspective, there is too much (i) active trading, (ii) trading on momentum, (iii) complacency about the long-term, (iv) excess volatility, (v) volatile volatility and (vi) premium for equity risk. While each puzzle has a simple behaviorist explanation—namely, that's how people are—the explanations aren't logically consistent. If people trade too hectically (i), why does trading on past trends (ii) appeal, unless it calms markets contrary to (iv) and (v)? If people greatly fear risk (vi), then why do they trade so much (i) given that trading is costly and excessively volatile (iv)? Alternatively, if they are complacent about the long-term (iii) but fear risk so much short-term (vi), why isn't the extra volatility (iv) relatively uniform contrary to (v)? Rational myopia provides coherent explanations for all these puzzles, which I will sketch below.

Active Trading: If capital markets behaved the way that orthodox finance typically models them, everyone would agree on fair price and its changes, leaving little cause for trading apart from resizing aggregate portfolios. Disagreements about risk regimes, or "heterogeneous beliefs" as economists say, need to be widespread to account for the massive trading we observe.

Models of heterogeneity in finance date to Hal Varian (1985) but haven't gained much traction. The problem is that disagreements about stable risks tend to fade over time. To resolve it, risks need to be unstable. Unstable trends breed confusion about current levels and likelihoods of change. Disagreements naturally wax and wane. Price dynamics for competitive markets of heterogeneous agents turn out to be analogous to belief dynamics for rational individuals.

Momentum Trading: In orthodox theory, there's no point to trading on past price momentum, since current price encapsulates all known information about future prices. This line of thinking is known as the efficient market hypothesis (EMH) and was first clearly expressed by Eugene Fama (1965), although it traces back to Louis Bachelier (1900).

Most practitioners demur. Antti Ilmanen (2011) found that trading on past price momentum worked more broadly and consistently than any other simple investment strategy. One of the world's top quant funds, AQR, was expressly built on momentum trading. Ironically, AQR founder Cliff Asness (2016) built his first models under Fama's tutelage. He credited Fama for being "incredibly supportive" although "it is obviously a result that he never really liked". Fama himself acknowledged that "momentum is a big embarrassment" to EMH.

The core problem with EMH is the inherent opacity of future risks. Wary of their own forecasts, traders tend to defer to market consensus and trim their bets. Smaller bets slow market adjustment which in turn breeds momentum trading as an offset.

Complacency about the long-term: The NPV of securities often hinges on payoffs a decade or more down the road. Yet capital markets rarely display much interest in long-term risks. Active traders focus on fast-changing news they can get a jump on. Visions of the future excite them only when they imagine other investors getting excited too. Risks of disaster tend to get little attention until disaster strikes. Reviewing centuries of evidence on sovereign debt servicing, Carmen Reinhart and Kenneth Rogoff (2009) concluded that lenders were too easily lulled into complacency.

This book offers a more innocent explanation. Rational, uncertain learners are bound to project recent good servicing into the future. On the bright side, this helps sovereign borrowers rebuild reputations after default and helps innovators gain backing. On the dark side, easy borrowing can help troubles mount and make a long-deferred reckoning worse.

Excess volatility: In finance, volatility refers to the annualized standard deviation of percentage price changes. Robert Shiller (2000) demonstrated that equity prices are far more volatile than dividends, whereas orthodox finance expects a near-match. He ascribed the difference to "irrational exuberance", the title of his book.

Rational myopia links excess volatility to uncertainty. Worrying that trends might have changed, traders put extra weight on current evidence. High earnings breed optimism about future growth; low earnings breed pessimism. Positive correlation between noisy earnings and noisy growth projections boosts price volatility.

Volatile volatility: The technical name is GARCH, which stands for Generalized Autoregressive Conditional Heteroskedasticity. Heteroskedasticity refers to variations in price volatility while autoregressive conditionality links the variations to past price history. Models of GARCH behavior, first developed by Robert Engle (1982), have proven extraordinarily useful.

However, GARCH is purely descriptive. While its guiding equations tell us how volatility behaves, they are disconnected from the NPV calculations that drive fair prices. At worst, GARCH models resemble the ad hoc layers of Ptolemaic epicycles. At best, they resemble Kepler's ellipses, focused correctly on the sun but with no link to gravity.

In fact, GARCH-type behavior is a rational response to uncertainty. Rational learners adjust fair prices roughly in proportion to the variance of their beliefs. Their variance responds to news roughly in proportion to a modified skewness of beliefs, which makes it volatile except in highly controlled conditions. Since skewness changes over time, the volatility of volatility tends to be even more irregular than GARCH models presume.

Premium for equity risk: Annual equity returns have long averaged on the order of 5 percent more than short-term debt returns. Rajnish Mehra and Edward Prescott (1985) contended that the observed risk premium on equities grossly exceeds what savers should reasonably need. Robert Barro (2006), elaborating a counterargument by Thomas Rietz (1988), showed that reasonable anticipation of rare consumption disasters can potentially justify the observed premium. Yet the implied risk aversion still seems unrealistically high. Moreover, the implied price behavior is strange when disasters are protracted.

Modeling the dominant market participants as long-term wealth-maximizers helps resolve this conundrum. Although their inherent risk aversion is low, risk premia must be high to induce them to hold securities with volatile excess volatility. In a sense, they must be paid to take the risk of other people's changing beliefs.

The next few chapters sketch several apparent dichotomies in finance and explain how a learning-based perspective can bridge them. Part 2 tracks rational learning in an unusual casino and shows that doubts about the stability of risk induce rational myopia. Part 3 extends this analysis to debt markets, where the odds of near-term servicing are nearly always very high. Part 4 addresses learning in equity markets, modeled as bets on the drift of random walks.

2

Risk versus Uncertainty

Informally, risk refers to the chance of a random outcome whereas uncertainty refers to our confusion over what the risk really is. To the extent we know the relevant risks, capital markets typically price them wisely. When we cannot know the risks, pricing often looks foolish. To analyze these phenomena, we need to distinguish objective probabilities that quantify risk from subjective probabilities that quantify uncertainty.

Risk is clearest when we can identify some symmetry in the randomness and deduce the probabilities from it. When we say that a jellybean is picked randomly from a jar, we mean that each jellybean has equal chance of being picked. If M jellybeans out of the total N beans are red, the probability θ of selecting a red jellybean is M/N. Deduction works superbly in physics, whose core laws are founded on symmetry across space and time. Deduction works well with coin tosses, dice rolls, roulette spins, and many card games. In most other applications, deduction fails to identify risk and we need to estimate it from the data.

Suppose we want to estimate θ in the previous example but can't count M and N. Instead we choose a bean randomly, record its color, return it to the jar, and repeat multiple times. Any assortment is possible unless θ takes the extreme value 0 or 1 and all randomness disappears. Yet we know that θ is more likely to lie close to the frequency of reds than far away. We can compare the perceived likelihood of various θ using subjective probabilities $p(\theta)$. Usually p is continuous, as we can't distinguish well between risks that are very close. Usually p has just one or two humps, which represent our leading hypotheses about θ. That still leaves p a lot of freedom, with far more variety than for θ itself. Figure 2.1 displays a few examples.

The various θ can be viewed as representing distinct hypotheses about the world, with $p(\theta)$ marking their relative credibility. Alternatively, we can describe θ as beliefs and $p(\theta)$ as their relative convictions, or just use beliefs as a catchword for both. The beliefs we bring to bear when analyzing new information are called prior beliefs or priors. Since people draw on different information and have different opinions on relevance, their priors tend to differ too.

Classical statisticians felt that uncertainty boded too much variation. They favored sticking with a baseline hypothesis until there was less than 5% chance or 1% chance that repeated independent, identically distributed ("iid") sampling would generate similar evidence. These aren't bad rules of thumb, as they set broad ranges of outcomes not to worry about and reserve alarm for outliers. For example, suppose a coin flips heads 58

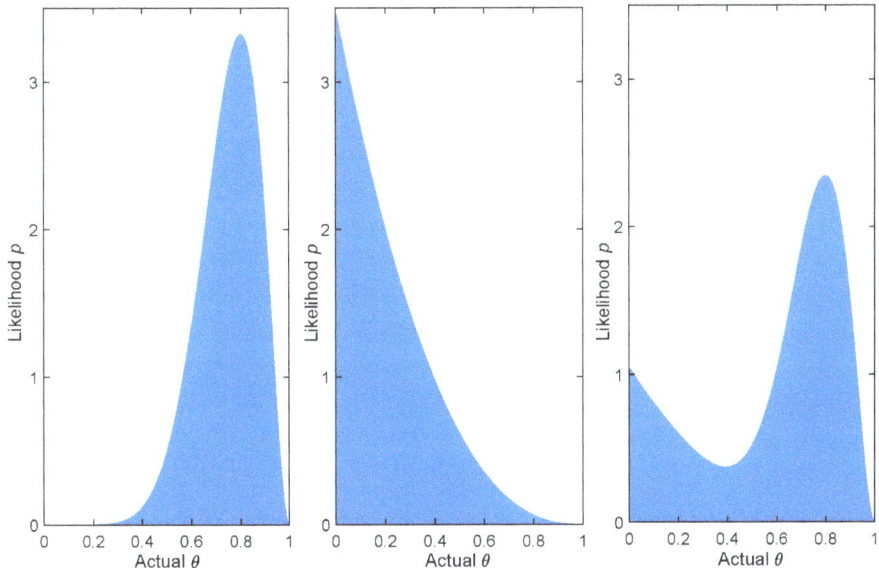

Figure 2.1: Possible uncertainties p about a single risk θ

times out of 100. With a baseline hypothesis of $\theta = 0.5$, the tightest 95% range or "confidence interval" spans 40 to 60 heads, as depicted in Figure 2.2, so the result is deemed consistent with the baseline.

Yet there ought to be better ways to respect gradual accretion of doubt. Ronald Fisher (1922) developed one, called maximum likelihood. It picks the risk distribution most likely to have generated the evidence given iid sampling. Figure 2.3 charts the probability of generating 58 heads for a host of different θ. Maximum likelihood picks $\theta = 0.58$ because that is where the graph peaks.

In many contexts, either approach is too confining. Classic confidence intervals promote stability but miss small changes. Maximum likelihood combs the data for new trends but is highly vulnerable to noise. How can we tap related evidence with more respect for doubt?

The best approach tracks uncertainty in more detail, along the lines first proposed by Thomas Bayes (1763). Like Fisher, Bayesians compute the various likelihoods $h(x|\theta)$ that x would occur if the risk were θ. Unlike Fisher, Bayesians don't narrow down to a single θ that maximizes $h(x|\theta)$.

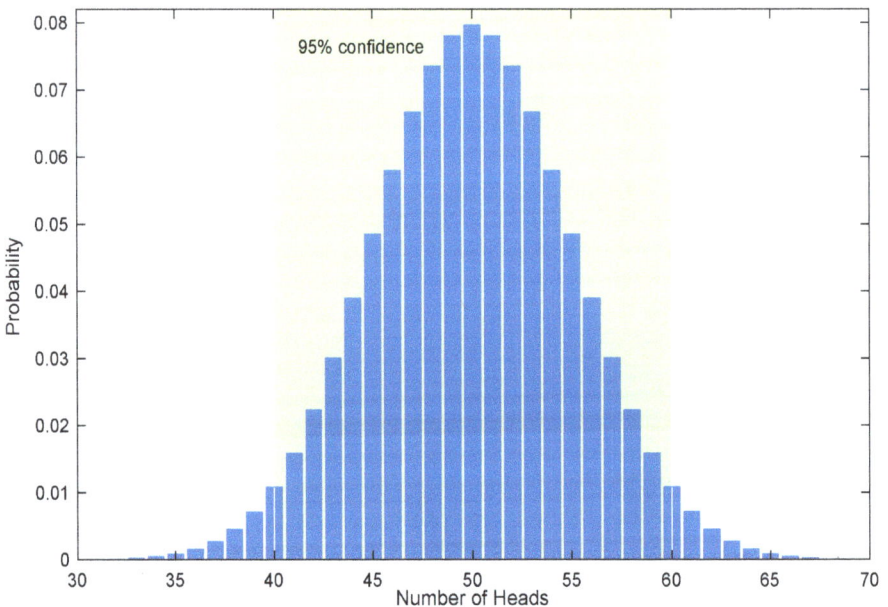

Figure 2.2: 95% confidence interval for heads in 100 flips when $\theta = 0.5$

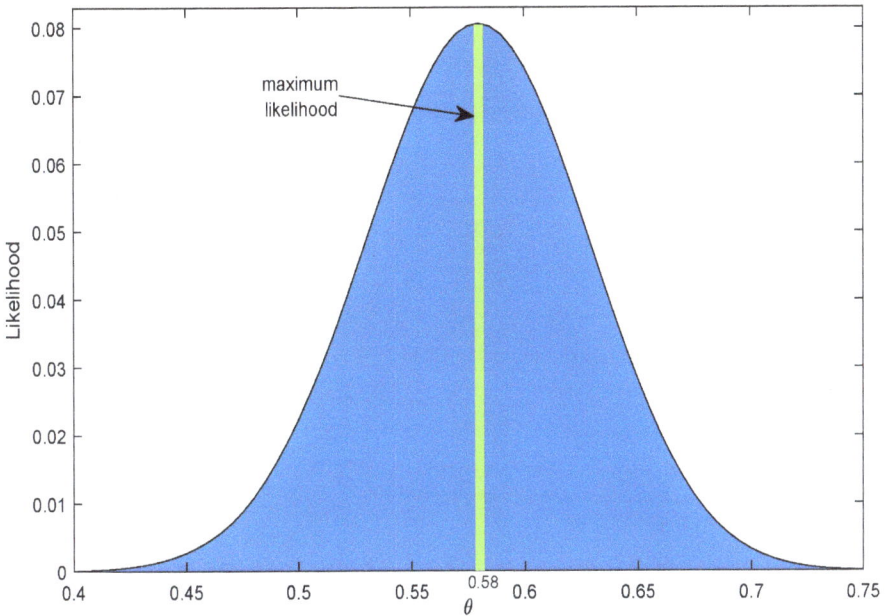

Figure 2.3: Likelihood of 58 heads in 100 flips for various θ

Instead, they reweight p in proportion to h:

$$p_{bayes}(\theta) \propto h(x|\theta) \cdot p(\theta). \tag{2.1}$$

This is known as Bayes' rule. It follows from the definition of conditional probability. To make the various p_{bayes} sum to one, divide the raw values by their sum or integral. For example, if initial beliefs were uniform between 0 and 1, p_{bayes} after 58 heads in 100 flips would be proportional to the likelihoods in Figure 2.3. If we started with more confidence that θ is close to 0.5, p_{bayes} would peak closer to the center and be more tightly distributed around it. While the computations can be tedious, priors taking a specified "conjugate" form allow quick shortcuts. Figure 2.4 displays an example when prior beliefs have an information impact equivalent to observing 50 heads in 100 previous flips.

In finance, hardly any key risks are replicated under clearly iid conditions, much less deducible from underlying symmetries. Instead, we

15

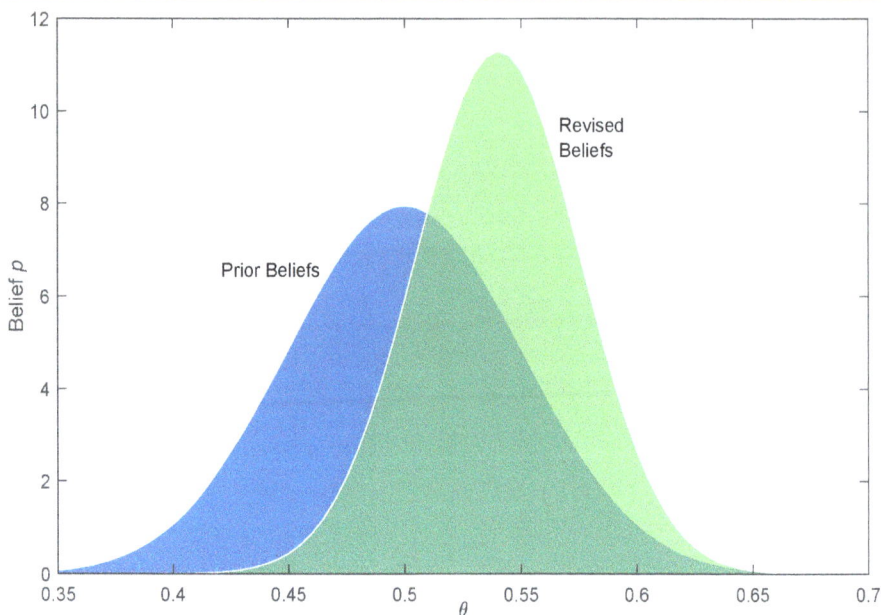

Figure 2.4: Bayesian updating for 58 heads in 100 flips when prior beliefs are bell-shaped with mean 0.5 and standard deviation 0.05

look for regularities in pools of partly similar assets or partly similar events. If our selection is sound enough and the pool large enough, we can identify common drivers of risk. However, even common drivers like average default rates or trend GDP growth can change dramatically in crisis. Absent such risks, a crisis wouldn't be much of a crisis.

Unfortunately, neither classical, maximum likelihood, nor shortcut-Bayesian methods work well when the common risks are highly unstable. Suppose we observe 50 straight heads when $\theta = 0.99$ followed by 50 straight tails when $\theta = 0.01$. While the aggregate is 50 heads out of 100, it is hugely misleading to suggest that $\theta \approx 0.5$. Here are some alternative methods for estimation:

- **Infer from 50 straight heads that their average $\theta \approx 1$ and infer from 50 straight tails that their average $\theta \approx 0$.** That works well in hindsight but doesn't assist prediction while the flips are being observed. By assumption, the coins don't advertise which bucket they come from.

- **Pick $\theta \approx 1$ if heads dominate in the previous three flips and $\theta \approx 0$ if tails dominate.** That's a great predictor for extreme values of θ, which it normally catches after just two observations. However, if fails miserably if θ switches back to a more central value.
- **Amend the application of Bayes' rule to allow for possible shifts in θ.** The results depend on the likelihoods assigned to different shifts, which might be little more than wild guesses.
- **Give up.** Declare the problem too idiosyncratic or ill-defined to warrant probabilistic analysis. Expect raw emotions or psychological quirks to prevail.

Uncertainty in Finance

"Give up" has a distinguished pedigree. Fisher rejected any use of probability to describe subjective beliefs. In finance, the two economists who most highlighted the uncertainty of capital markets ruled out its quantification. In *Risk, Uncertainty and Profit*, Frank Knight (1921) defined uncertainty as unmeasurable risk:

> A *measurable* uncertainty, or "risk" proper, as we shall use the term, is so far different from an *unmeasurable* one that it is not in effect an uncertainty at all. We shall accordingly restrict the term "uncertainty" to cases of the non-quantitative type. It is this "true" uncertainty, and not risk, as has been argued, which forms the basis of a valid theory of profit and accounts for the divergence between actual and theoretical competition. (Chapter 1)

> Take as an illustration any typical business decision.... What is the "probability" of error (strictly, of any assigned degree of error) in the judgment? It is manifestly meaningless to speak of either calculating such a probability a priori or of determining it empirically by studying a large number of instances.... [T]he "instance" in question is so entirely unique that there are no others [...] to form a basis for any inference of value about any real probability in the case we are interested in. (Chapter 7)

John Maynard Keynes thought along similar lines. In *A Treatise on Probability Theory* (1921), he distinguished beliefs that deserved probabilistic quantification from beliefs that did not. His work in economics placed most beliefs about capital squarely in the latter camp, where emotion and animal spirits were bound to trump rational calculation:

> The outstanding fact is the extreme precariousness of the basis of knowledge on which our estimates of prospective yield have to be made…. [O]ur existing knowledge does not provide a sufficient basis for a calculated mathematical expectation. (Keynes 1936, Chapter 12)

> By "uncertain" knowledge, let me explain, I do not mean merely to distinguish what is known for certain from what is only probable. The game of roulette is not subject, in this sense, to uncertainty…. Even the weather is only moderately uncertain. The sense in which I am using the term is that in which the prospect of a European war is uncertain, or the price of copper and the rate of interest twenty years hence…. About these matters there is no scientific basis on which to form any calculable probability whatever. We simply do not know. (Keynes 1937)

In short, Knight and Keynes thought that the instability of major financial risk and the dearth of iid observations ruled out probabilistic analysis. For the probability theory they knew, they were correct. They weren't aware of Markov processes, which allow changing risks to exhibit long-term stationarities. Notions of martingales—symmetrically fair games with potentially never-ending payoffs—were just being developed.

Nevertheless, Knight and Keynes were wrong to claim that uncertainty can't be quantified. Imagine a lottery ticket paying one if a specified event occurs and zero if it doesn't. The most we'll purchase the ticket for is our mean belief that the event occurs. That simple linkage, first noted by Bruno de Finetti (1937) and elaborated by Leonard Savage (1954, 1971), can be extended to elicit expected quantiles or higher moments over a continuum of possible outcomes.

By the 1960s, subjective probability was more accepted in economics. In a now-classic article "Proof that properly anticipated prices fluctuate randomly". Paul Samuelson contended that efficient capital markets incorporate all available information, including agents' probabilistic assessments of future returns. However, he did not claim that markets predicted the future well and acknowledged that his treatment raised a host of new questions:

> I have not here discussed where the basic probability distributions are supposed to come from. In whose minds are they ex ante? Is there any ex post validation of them? Are they supposed to belong to the market as a whole? And what does that mean? Are they sup-posed to belong to the 'representative individual', and who is he? Are they some defensible or necessitous compromise of divergent expectation patterns? Do price quotations somehow produce a Pareto-optimal configuration of ex ante subjective probabilities? This paper has not attempted to pronounce on these interesting questions. (Samuelson 1965)

The next generation of theorists focused on the neatest subset of these questions: "rational expectations" equilibria where trends are fore-cast accurately. While this provided invaluable insights, it glossed over the challenges of forecasting trends that aren't stable. Imagine that your dog, carrying risks, breaks loose and that you, carrying beliefs, try to catch it. If the dog maintains a steady path, you steadily approach and recapture it, reuniting beliefs with risks. However, if the dog keeps changing direction, the equilibrium behavior is chase rather than capture, and beliefs will occasionally diverge from risk.

Albert Einstein famously rejected quantum theory on disbelief that God would play dice with the universe. In finance, it seems that God occasionally changes the dice without telling us. Capital markets cannot foresee the odds. At best, they're great at learning from experience. I will call this constrained optimum the Efficient Learning Hypothesis or ELH. Efficient learning captures the spirit of what Mordecai Kurz (1994ab, 1997, 2009) called "rational beliefs" but focuses more on process.

To apply ELH, we need to amend the application of Bayes' Rule. It is not enough to update beliefs rationally about θ's value in the latest round of observation. To anticipate possible change in θ at the next round, (2.1) should be supplemented with

$$p_{new}(\theta) = p_{bayes}(\theta) + \langle \text{net migration into } \theta \rangle, \tag{2.2}$$

where $\langle \; \rangle$ denotes the expected value given beliefs p. The strong form of ELH asserts that learners understand the migration process. In effect, strong ELH is rational expectations constrained only by the objective risks in future trends. In practice, it is even harder to gauge the migration process than to track the current θ. We often make crude simplifying assumptions to ease computation. This replaces the equality in (2.2) with approximation, leaving a weak form of ELH. Weak ELH says that people use rules of thumb to approximate efficient learning.

A good example of weak ELH is an exponentially weighted moving average or EMA. It nudges the current forecast E a constant fraction η, known as the learning rate, toward the latest observation x. At first glance, there's nothing probabilistic about it: just one number E that keeps shifting. Nevertheless, EMAs can be viewed as a reduced form of Bayesian updating, where migration stabilizes the variance of beliefs.

An EMA can be an excellent predictor when x is a random walk with gradually changing drift, provided η is set appropriately. If η is too high, the EMA will bounce around way too much. If η is too low, the EMA will lag far behind changes. If we don't know the appropriate η, we can start with a wide range of EMAs and let Bayesian updates select over time for the better ones. We can also allow for the best EMA to change over time, which in effect imposes another layer of (2.2). As we will see, the layering tends to correct errors. However, there is always a tradeoff between learning speed and turmoil.

3

Calm versus Turbulence

Since no one can sample the future, reliable forecasting hinges on identifying stable laws of motion and a few dominating influences. Future weather is hard to predict as there are too many drivers with too uncertain pace. Future returns on capital are even more opaque. So much depends on products that haven't yet been invented, rivals that haven't yet emerged, regulations that haven't yet been imposed, and disasters that haven't yet struck.

While the capital market anticipates future rewards and risks, it can't anticipate them the way a halfback racing down the field anticipates the reactions of defenders. The market never sees clearly what lies in store. Instead it observes the profits and perils it weaved through before and guesstimates what might come next. In short, it looks backward into the future.

Backward into the future? In most languages that sounds bizarre. Words like "ahead" and "forward" that point to space in front of us also point to future time. However, that isn't the only way to align directions. A handful of languages link by what we observe. The most prominent is Quechua, the Incan language still spoken by more than 6 million people. In Quechua, *ñawpa* means both "ahead" in space and "past" in time while *qhipa* means both "behind" and "future".[2]

Looking backward into the future is useful since the future always recapitulates some of the past and usually recapitulates most. Surprises usually are outliers from a still intact trend. We tweak the next round of forecasts and move on. Occasionally the surprises get under our skin. They shock us less by their size than their persistence. They make us suspect that what we thought of as a rare outlier is now the new norm.

Responses to shocks are rarely even. Price charts typically exhibit three phases: muted response, a huge surge in response and return to normalcy. The market appears to flip from complacency to overreaction and then gradually regain its bearings. Yet there is nothing inherently irrational about it.

Let me demonstrate this with the help of a peculiar coin and a rational, disembodied robot ("bot") bred to predict it. The bot knows that the chance θ of heads takes only two possible values, 0.9 or 0.5, with convictions $p \equiv p(0.9)$ and $1-p$ respectively. Bayesian updates boil down to the following simple rule: add $\log(0.8) \approx 0.58$ to the log odds $\log(\frac{p}{1-p})$ if heads and subtract $\log(5) \approx 1.61$ from the log odds if tails. However, the way markets price risk obscures that. Price mostly reflects the bot's mean belief $E = 0.5(1-p) + 0.9p$, which indicates the expected value of a payout of 1 on heads and 0 on tails. That is linear in p but not in log odds.

For our first simulation, suppose that p starts out extremely low, just one in a million, and that the next 60 flips alternate nine heads and one tail. Figure 3.1 charts the evolution of E as a solid line. The response is very mild at first, as if the bot is dozing. E doesn't breach 0.51 until the 25th flip, Suddenly, E gets ultra-sensitive, as if the bot awakens in panic. Over the next 20 flips E reaches 0.89, with five straight heads adding 0.25 and two tails deducting over 0.07 each. A few plays later, calm is restored. The tail at flip 50 prompts barely more reduction than the tail at flip 20.

How special is the initial p or the timing of tails? Not very. Multiplying p by 10 would start the surge about four observations sooner but keep the core transition similarly quick. Making tails random would alter the size and timing of interim setbacks but not the dominant S-shape.

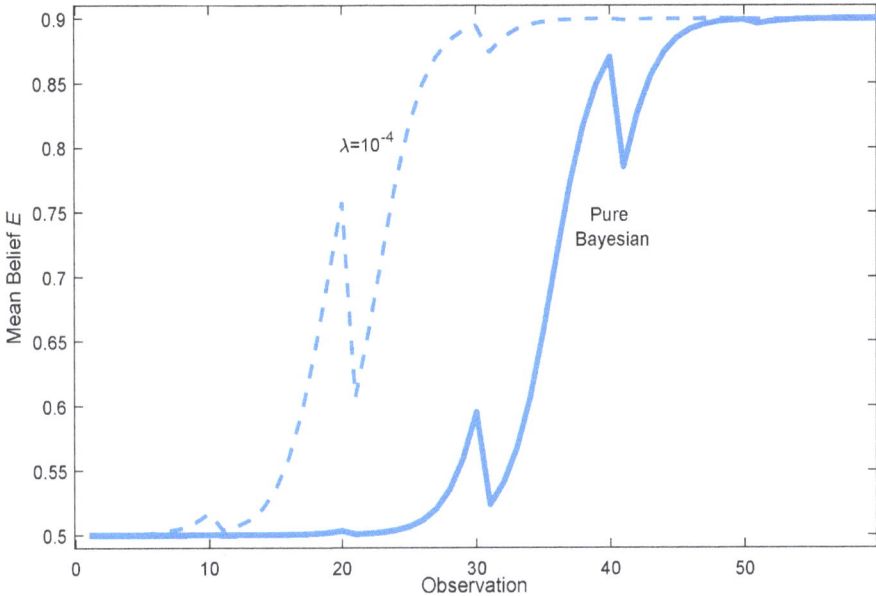

Figure 3.1: Mean belief before every observation when $\theta = \{0.5, 0.9\}$, initial $p(0.9) = 10^{-6}$, and nine heads alternate with one tail

Is the bot sensible to react this way? Not completely. Its Bayesian updates presume θ never changes. An alternative bot allows a small switching probability of $\lambda = 10^{-4}$. Applying (2.2), we add $\lambda(1-2p)$ to p after every Bayesian update. Since this keeps p from falling below λ, the surge starts sooner, without making the core transition less steep or less volatile. Figure 3.1 charts its mean belief as a dashed line.

In machine learning, most algorithms using observations x to update a mean E have a core component $\Delta E = \eta(x - E)$, where Δ denotes the change, $x - E$ is the element of surprise in the news and η is a small positive fraction known as the learning rate. I will invert this relationship to define $\eta \equiv \Delta E / (x - E)$. Figure 3.2 charts the learning rate for the dashed line in Figure 3.1 and shows that it soars during the core transition. To emphasize how weird this is, Figure 3.2 also charts the reciprocal $T_{eff} \equiv 1/\eta$ on a log scale. I call T_{eff} the effective sample size, since it indicates the number of iid-equivalent observations folded into E.

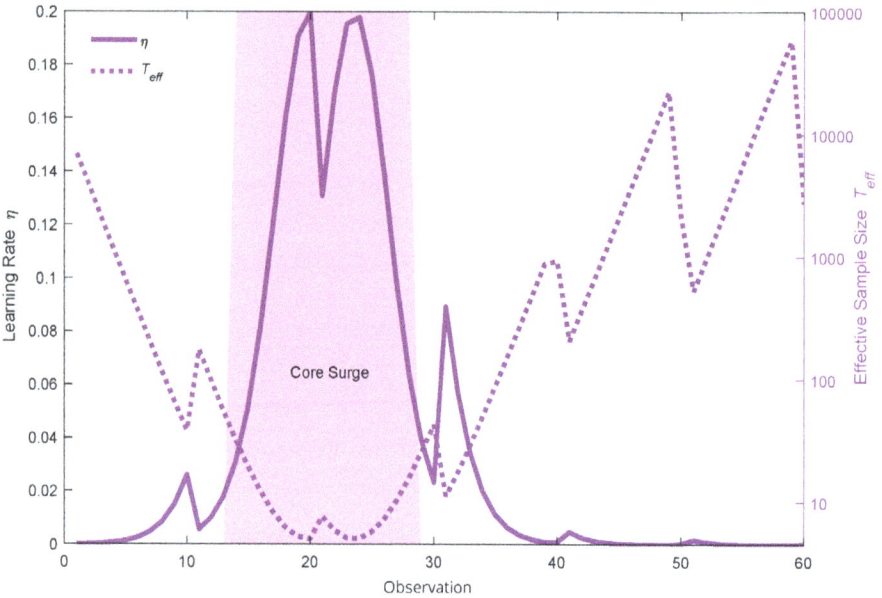

Figure 3.2: Learning rate and effective sample size for the dashed line in Figure 3.1

For example, suppose a head raises E to 0.8 from 0.75. The implied $T_{eff} = 5$ makes this equivalent to starting with four iid observations–three heads and one tail—and adding one more iid head to the mix. While T_{eff} exceeds 500 for most of the observations, during the peak surge it averages less than 10. The minimum occurs after observing the 22nd head in the past 24 flips. Presumably the bot compares those flips to a much longer sequence reassuring that θ is still 0.5. Why then isn't T_{eff} at least 24? Because the bot isn't confident that the previous mean is relevant.

Nassim Taleb (2001, 2007) argued that people are often "fooled by randomness" into treating the small, ordinary risks of "Mediocristan" as the only risks they will ever experience. In "Extremistan", where unusual "black swan" risks strike, that works badly. At first people respond too complacently and later they panic, with market prices mirroring their reactions. The previous simulations show that rational bots can seem just as foolish. When the Mediocre $\theta = 0.5$ switches to an Extreme $\theta = 0.9$, the solid line in Figure 3.1 applies a standard Bayesian update each round yet seems to flip from complacency to panic. The dashed line allows for change in θ every round, which quickens response and seems to panic even more.

This demonstration challenges the notion that panic is the prime cause of market turbulence. Perhaps turbulence stirs fear through the high risks it entails, which in turn triggers panic. If so, market turbulence will resemble aircraft turbulence. In fact, there is a deeper kinship between the two types of turbulence. Like physical fluids, beliefs tend to flow in ways that conform with their containers. In capital markets, the flow is called learning and the container is history. Slow learning is reasonably calm. Very fast learning causes rational turbulence. What makes learning slow? Evidence that history is largely repeating itself. We're refining our estimated values, not changing them drastically. What makes learning fast? Evidence that past norms are no longer relevant.

Since Figure 1.2 parodied academic views, I will let Figure 3.3 parody mine. People in the tower are viewing calm evidence but heading toward a turbulent future, so I need to temporarily reverse the laws of submarine propulsion. I will restore them when the tower reviews a turbulent past.

Figure 3.3: How finance "ought" to view water

Cumulant Hierarchies

The distinction between calm and chaotic market phases dates to Benoit Mandelbrot (1963). He noted that small changes in market prices tend to be followed by small changes and big changes by big changes. The observations helped drive his discovery of fractals. Later Mandelbrot (1997) likened market price movements to physical turbulence and used fractals to illustrate their common connection. However, no fractal description can identify the key drivers, much less distinguish the rational and irrational components.

To connect rational beliefs with material fluids, I will elucidate a common feature in the equations driving their flow. Called a "cumulant hierarchy", it induces cascades that can either smooth random fluctuations or amplify them. Since cumulant hierarchies aren't easy to grasp and typically lack neat solutions, let me try to make them more palatable by sketching the context of their discovery.

Modern physics traces both smooth and turbulent flows to a single differential equation known as Navier-Stokes (Navier 1822, Stokes 1851). It equates the aggregate force on each portion of fluid to its mass times its acceleration. Conservation of momentum ties flow nonlinearly to changes in flow. A key driver is the ratio of momentum to internal friction (viscosity), known today as the Reynolds number (Reynolds 1883). Low Reynolds numbers are associated with smooth flows. High Reynolds numbers trigger turbulence.

Despite Navier-Stokes' one-line simplicity, why it induces occasional turbulence it does is hard to figure out. The mathematical physicist Horace Lamb allegedly joked that

> when I die and go to heaven, there are two matters on which I hope for enlightenment. One is quantum electrodynamics and the other is the turbulent motion of fluids. And about the former I am rather optimistic. (Lamb 1932)

Richard Feynman, who illuminated the perplexities of quantum electrodynamics, called turbulence the most important unsolved problem of classical physics:

> [T]here is a physical problem that is common to many fields, that is very old, and that has not been solved.... Nobody in physics has really been able to analyze it mathematically satisfactorily in spite of its importance to the sister sciences. It is the analysis of circulating or turbulent fluids. (Feynman 1963, Chapter 3)

No one has managed to model turbulent flow in a way that fails to diverge over space and time. Even supercomputers, which update Navier-Stokes over massively detailed grids, can't yet predict major storm paths reliably for more than a few days at a time. Mathematically, each moment, which is the mean of a power function x^n, depends on the moment one power higher. According to an esteemed textbook,

> The problem of closing the moment hierarchy is usually referred to as the "closure problem" and is the underlying problem of turbulence theory. (McComb 1990)

The effect is clearest when we shift to cumulants, which are moments adjusted for the impact of lower moments. The first cumulant κ_1 is the mean E, the second cumulant κ_2 is the variance V, the third cumulant κ_3 is directly proportional to the skewness, and the fourth cumulant κ_4 is directly proportional to the (excess) kurtosis. When the higher cumulants dissipate, flows are smooth. When higher cumulants surge, they resonate all the way down and create turbulence.

Why is the closure problem fundamental to turbulence? Imagine two drops of fluid nearby each other in space or time. As the surrounding fluid distributions and forces are very similar, only very high cumulants will register any noticeable differences. Nevertheless, each cumulant impacts the cumulant below, which given sufficient stress can make the two drops behave very differently. That's the essence of turbulence. Reduce the stress and the higher cumulants shrink toward the zeros that characterize a Gaussian distribution. That is, turbulence subsides if and only if random variations cancel out more meaningful patterns in the noise.[3]

Rational filtering of evidence is also governed by a single equation and it too generates a cumulant hierarchy. The Core Learning equation takes the form

$$\Delta p(\theta) = p(\theta)(\theta - E)\frac{x - E}{\sigma_E^2} + \langle \text{net migration into } \theta \rangle, \tag{3.1}$$

where σ_E^2 denotes the perceived variance of x conditional on E. Remarkably, it works for nearly any beliefs for nearly any random process we encounter in finance, provided we take limits in continuous time wherever possible.

Bear in mind that (3.1) applies to every feasible θ at every time, which in the limit demands infinitely many updates infinitely often. Focusing on aggregates like E helps trim updates to manageable size. Multiplying both sides of (3.1) by θ and integrating, we find that

$$\Delta E = V\frac{x - E}{\sigma_E^2} + \langle \text{migration-induced shift in } E \rangle. \tag{3.2}$$

This is simpler than (3.1) and easier to interpret. The greater the variance in beliefs relative to the variance in outcomes, the greater the response to news. Generalization proves that all neighboring cumulants satisfy

$$\text{volatility}(\kappa_n) \approx \frac{|\kappa_{n+1}|}{\sigma_E} \tag{3.3}$$

where $| \; |$ denotes absolute value. Equation (3.3) gives a crisp expression of cumulant hierarchy. The more Gaussian the beliefs, the smoother the associated learning. Conversely, if extreme outliers emerge and repeat, they will boost higher cumulants and resonate below. Turbulence will persist until E aligns more closely with the true mean of x.

Lars Hansen (2007) acknowledged the practical challenge in updating the distribution of beliefs, which "for many state spaces can be an infinite dimensional state variable" and in principle should be updated every infinitesimal instant. However, Hansen declared this "a computational issue, not a conceptual one". Rewriting the optimal updating rule in terms of cumulants shows that Hansen was mistaken. Updating a never-ending cumulant hierarchy is indeed a conceptual issue, since it explains why beliefs converge in some contexts and diverge in others.

Since most capital market prices are closely related to mean beliefs, (3.2) relates the variance of market prices to the variance of beliefs, which from (3.3) can be highly volatile too. Hence, the cumulant hierarchy offers a neat, rational explanation of GARCH behavior. The cumulant hierarchy also explains why disagreements ebb and flow so much in capital markets. When risk shows signs of shifting a lot, tiny doubts are bound to inflate disagreements. Conversely, stability deflates disagreements. Standard models miss this because they disregard all potential instability and associated doubts. Without that, calming influences dominate (3.3) and shrink beliefs toward the mean.

It doesn't take much doubt to sow rational turbulence. Bear in mind that $(\theta - E)/\sigma_E$ measures the relative idiosyncrasy of a belief θ while $(x - E)/\sigma_E$ measures the relative surprise in the news. Hence the Bayesian component of (3.1) indicates that:

$$\% \text{ change in conviction} = \text{idiosyncrasy} \times \text{surprise} .$$

Convictions on big idiosyncrasies gain the most from surprises that tilt their way and lose the most from surprises that tilt against. Balanced surprises favor moderate views that converge calmly on the mean. Imbalanced surprises favor extremes and breed turbulence. However, even big percentage changes won't matter if the starting convictions are infinitesimal. Maintaining tiny doubts sets a floor under convictions. While they don't predict a big shift, they keep us open to the possibility that one has recently occurred.

4

Mind versus Market

The impact of cumulant hierarchies is best appreciated through simulations. Figure 4.1 charts the risk θ underlying 100,000 consecutive coin flips. Usually θ stays the same from one flip to the next. However, each period allows a small chance $\Lambda = 0.001$ of reset. When reset occurs, θ is resampled with every value between 0 and 1 equally likely.

Figure 4.1: Sample path for risk given probability $\Lambda = 0.001$ each period of resampling from uniform distribution

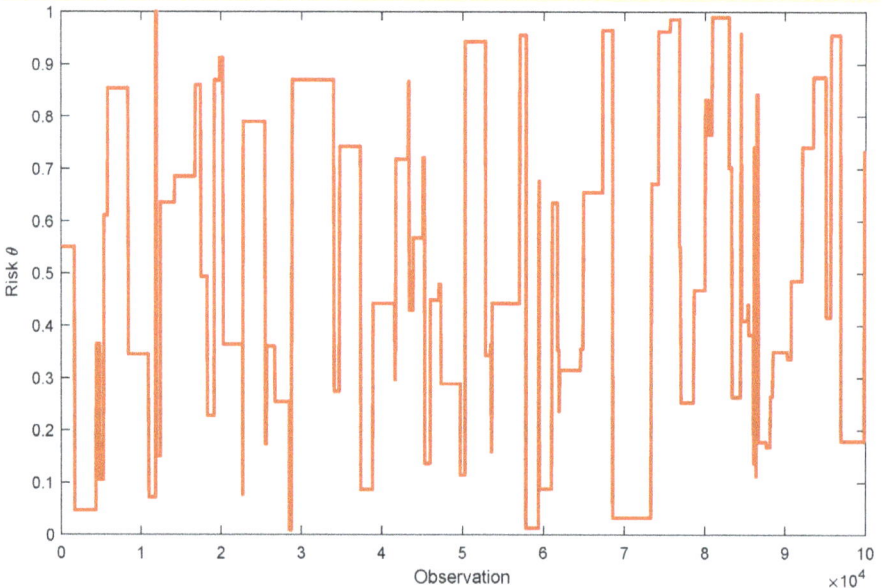

Bear in mind that no market participants directly observe θ. They see only a long sequence of heads $x = 1$ and tails $x = 0$, with each x sampled at the then-prevailing risk. Let's hire a bot to analyze the data and try to discern the underlying θ. Luckily, a bot named Perfect Learner (PL) just applied for the job. I wonder how it justifies its name.

PL: *For starters, I observe every outcome correctly.*

KO: So does everyone else in this book.

PL: *I also process all evidence completely rationally.*

KO: As do many bots. Do you observe θ?

PL: *No. But I do know exactly how it gets picked, namely, random resampling at rate* $\Lambda = 0.001$ *from a uniform distribution.*

KO: Amazing. How do you know?

PL: *I just do.*

KO: I don't get it. Even if you studied millions of coin flips prior, you can't be sure you nailed Λ. Perhaps Λ just changed.

PL: *No, it didn't. I know* Λ *because I'm a Perfect Learner. Now stop messing with my virtual head and let me study. There, done already.*

Figure 4.2 charts PL's mean beliefs heading into each coin flip. Most of its predictions are good, with 70% lying within 0.02 of the true θ. Yet there's also a lot of turbulence.

Figure 4.3 displays a close-up of 10,000 observations, this time with E overlaid on θ. We see numerous spikes, some of them tall, where PL briefly gets fooled by randomness. There are also stretches of over 100 observations where PL seems slow to adjust.

While perfect learning is far more realistic than perfect foresight of θ, it too exaggerates our acumen. Even under strong ELH, learners will at best estimate a range of likely Λ and resampling distributions with some allowance for possible change. The extra uncertainty will aggravate the skittishness evident in Figure 4.3.

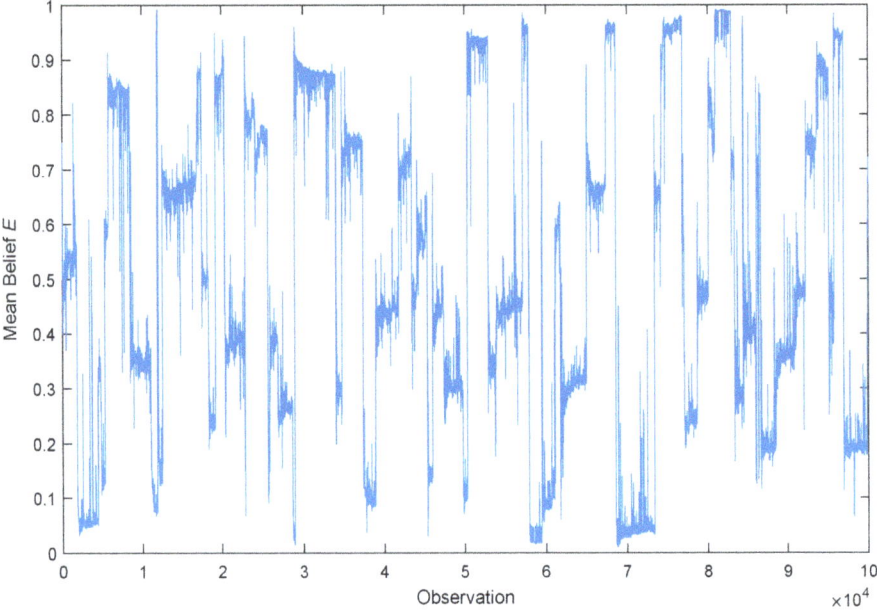

Figure 4.2: Mean beliefs for Figure 4.1 given perfect learning

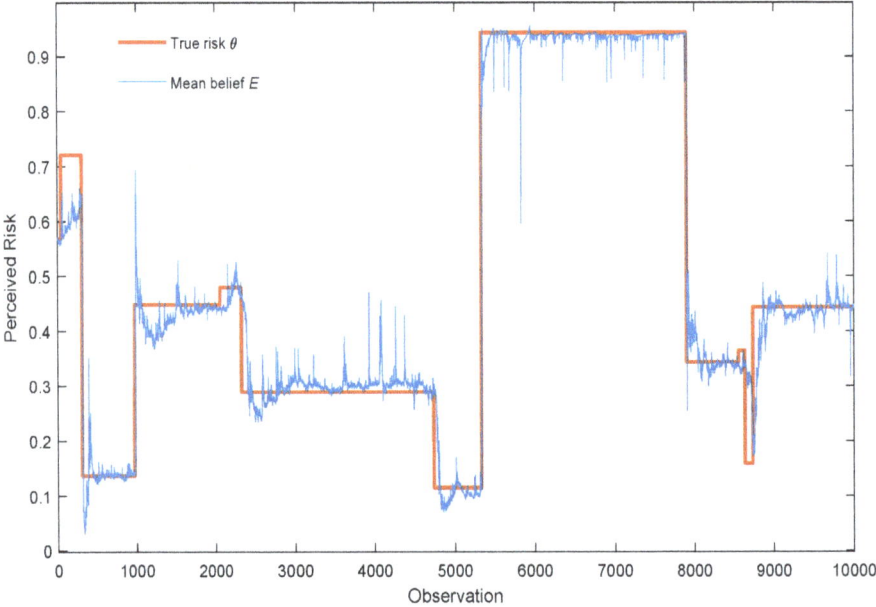

Figure 4.3: True risks and mean beliefs for 10,000 observations from Figures 4.1 and 4.2

Weak ELH invokes even coarser approximations. Consider EMAs with constant effective sample sizes T_{eff}. Figure 4.4 plots three examples for the observations underlying Figure 4.3. Compared to PL, all the predictors are lousy: too slow-moving, too jittery, or a mixture of both.

Despite these shortcomings, EMAs are useful building blocks for something that performs much better. Suppose we create 100 different EMAs, each with a different T_{eff}, and assign a relatively uniform prior to express our ignorance. Bayesian updates will soon select for the better EMAs. Given $\Lambda = 0.001$, the best T_{eff} is close to 33. As Figure 4.5 shows, its EMA jitters much less than the EMA for $T_{eff} = 10$ does and adjusts to new θ much faster than the EMA for $T_{eff} = 100$.

The best single EMA still doesn't look nearly as good as PL in Figure 4.3. Fortunately, we don't need to settle on a single EMA. Suppose we incorporate doubt over which EMA is currently best, modeled as some fixed rate λ of reset. Since we don't know which λ to choose, let's try a dozen different values, generate a different aggregate predictor for each, and add another layer of doubt to help choose between them. I spaced the

Figure 4.4: Mean beliefs for various EMAs given observations for Figure 4.3

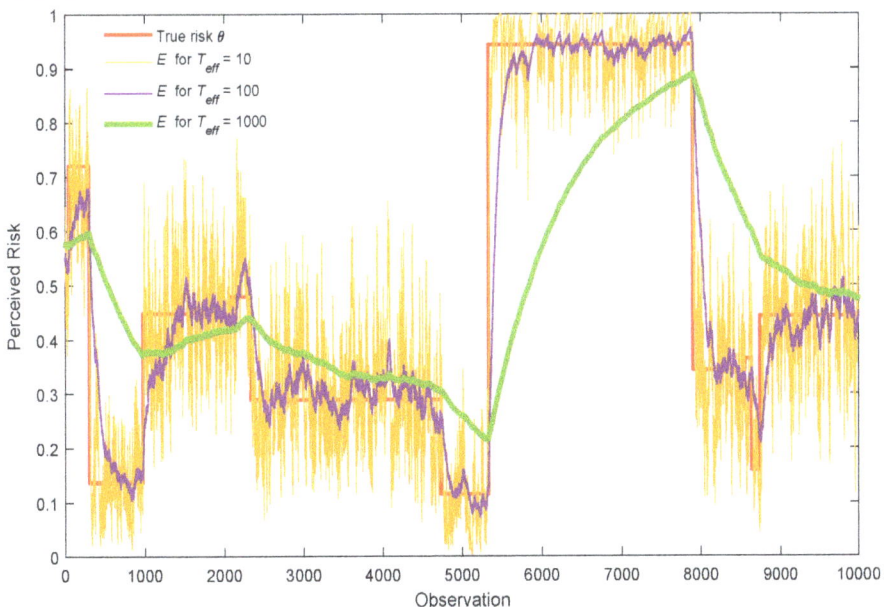

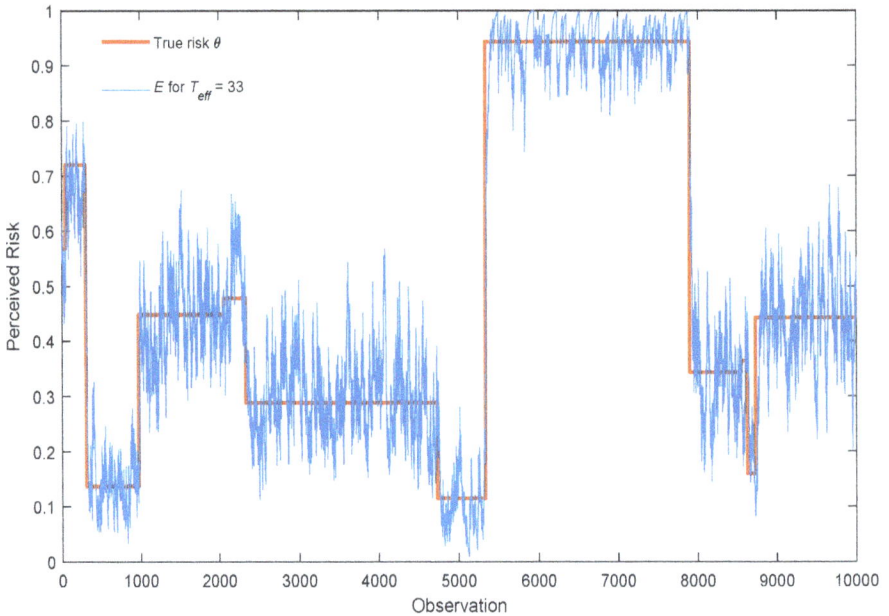

Figure 4.5: Sample path for the best single EMA predictor

T_{eff} of 50 EMAs geometrically between 2 and 5000, spaced the dozen doubts λ geometrically between 10^{-1} and 10^{-8}, and set the top-level doubt at 10^{-5}. However, the core results displayed in Figure 4.6 are highly robust to variation.

This greatly improves the fit. Nearly a third of the time, the dynamic mix of EMAs predicts θ even better than PL does. While PL's predictions have slightly lower mean squared error overall, comparing Figures 4.3 and 4.6 wouldn't prove that with 95% confidence; the sample would need to be nearly twice as long. How about comparing the daily changes in E? PL's predictions usually change more smoothly but exhibit occasional wild spikes. This makes the EMA-mixing bot seem more consistently rational: less likely than PL to get lulled into complacency or startled into panic.

Here is another way to layer simple bots. On the bottom layer, employ 21 dogmatists, each of them convinced that θ is an integer multiple of 0.05 and that every other dogmatist has the multiple wrong. On a second layer, employ a dozen screeners who rotate rationally among the dogmatists, with each screener applying a different doubt λ. On the top layer, have a

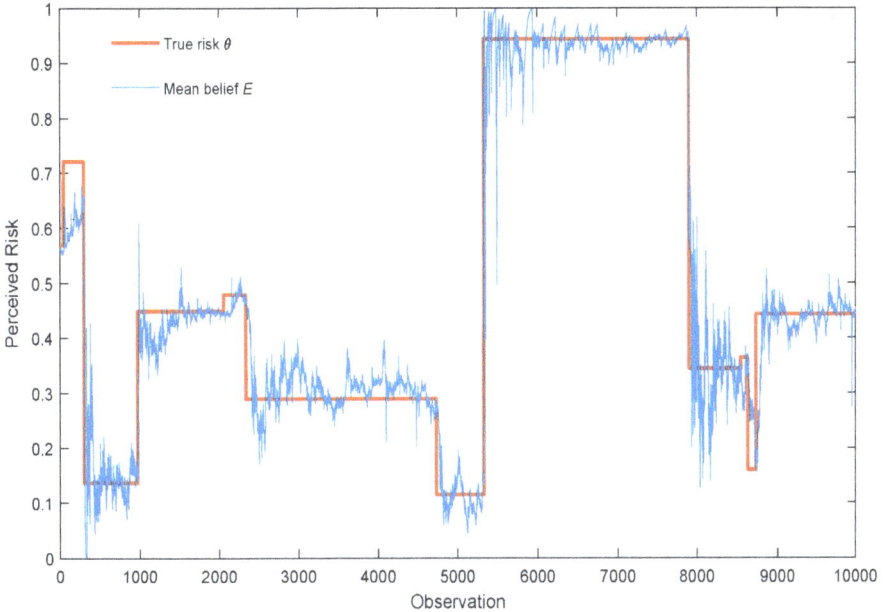

Figure 4.6: Sample path for a self-adjusting mix of EMAs

supervisor use standard Bayesian updates, with hardly any doubt, to focus on whichever screener rotates best. Despite the bots' complete ignorance of the underlying process, the tracking is so close to perfect learning that I won't bother to display it separately.

Market Mechanisms

Real-life capital markets entwine investors who are far less thorough or consistent than any of the bots described above. Do markets tend to (a) aggravate the shortcomings of its participants, (b) mitigate individual shortcomings but add collective distortions, or (c) make learning more efficient? The answer is (c). At their best, markets operate like an optimal individual learner. While participants don't share much information, market prices tell them most of what they need to know. Compressing others' thinking down to price also clears out informational clutter. It helps each trader focus on buy and sell orders, which in turns helps to refine price.

What makes this work is a gambling strategy intended to maximize long-term growth of capital. It is known as the Kelly criterion or more colloquially as Fortune's Formula (Kelly 1956, Thorp 1960, Poundstone 2005). It advises traders to size independent binary or Brownian bets in direct proportions to their gambling capital and their expected profit per bet and in inverse proportion to the conditional variance per bet. It also advises that the best proportionality coefficient on the combination is one:

$$\text{Optimal Size of Bet} = \text{Capital} \times \frac{\text{Expected Profit}}{\text{Conditional Variance}} \quad (4.1)$$

Unlike Bayesian updating, the Kelly criterion (4.1) isn't a must-do for rationality. A gambler who bets more can expect higher average profits. A gambler who bets less suffers fewer losses. However, the Kelly criterion maximizes the expected log growth rate of wealth and consequently is almost certain to outgrow any other strategy long-term. We will confirm this in Chapter 9 for binary bets and Chapter 18 for Brownian bets.

Suppose a security paying x is tentatively offered at price F with conditional variance σ_F^2. Suppose gambler bot i, whose share of the total capital K is k_i, thinks the mean is E_i. Applying (4.1), it aims to own $k_i K(E_i - F)/\sigma_F^2$ securities, where a negative number indicates a short position. For supply to match demand, price must equal the capital-weighted mean belief. Since the net payoff to each security is $x - F$, the outcome changes each capital share by

$$\Delta k_i = k_i \left(E_i - F \right) \frac{x - F}{\sigma_F^2} \quad \text{where } F = \sum_i k_i E_i . \quad (4.2)$$

If we replace E_i with θ_i, the capital share k_i with the subjective probability $p(\theta_i)$ and F with the mean aggregate belief $E \equiv \langle \theta_i \rangle$, (4.2) matches the Bayesian component of the Core Learning equation (3.1). As for the net migration component, we can replicate that in several different ways. For example, have each bot j own a share of the capital that bot i manages, couple wealth taxes with lump-sum transfers, or treat each bot i as a large pool of individuals who defect probabilistically to other pools.

In short, the Core Learning equation applies just as well to a fully Kelly-driven market as it does to a rational learner. Even though traders compete vigorously against each other and share no information other than the bids that determine price, the market behaves like an optimally unified rational learner. I call this the Invisible Mind theorem.

The isomorphism recalls the theoretical equivalence between market economies and centrally planned economies. In practice, central planning is far too cumbersome and too limited in the information it can handle to effectively supersede markets. Similarly, a decentralized capital market tends to learn significantly faster than individual participants. Its main advantages are the breadth of information it digests and its speed of digestion. The invisible mind that steers capital markets complements what Adam Smith described as the invisible hand that steers production.

However, capital markets are not fully Kelly-driven. Suppose my research suggests that the fair price is E while market consensus sets a fair price for F. Out of respect for the consensus, I might accord it a $1-f$ chance of being correct. This shaves my bet to a fraction f of full Kelly, which if applied consistently is known as fractional Kelly. Fractional Kelly can be viewed as a type of herding. While the term is often used pejoratively to suggest that people are too dumb or lazy to think for themselves, it is hard to fault some deference to the wisdom of the crowd.

Fractional Kelly can also stem from risk aversion. Full Kelly gambling is gut-wrenching. William Chin and Marc Ingenoso (2007) derived the chances of various drawdowns (maximum percentage losses) given a long sequence of independent Kelly-driven bets. For full Kelly, the probability of at least temporarily losing a fraction L of wealth is $1-L$, e.g. a 75% chance of a 25% drawdown and a 25% chance of a 75% drawdown. Big drawdowns sap convictions and prod losers to bet wildly or throw in the towel. To reduce those risks, most professional gamblers opt for fractional Kelly, typically with f of one-half to one-quarter (Chin and Ingenoso 2007). While half Kelly cuts the expected growth rate of wealth by 25%, it trims the probability of a 75% drawdown to less than 2%. While quarter Kelly cuts the expected growth rate by 56%, it trims the probability of a 50% drawdown to less than 1%.

To the extent that people gamble more cautiously than full Kelly advises, it seems unfair to describe them as "irrationally exuberant" or "fooled by randomness". Perhaps they so rue those tendencies that they veer the other way. Perhaps they become too hesitant to bet their own convictions and too fearful of being fooled.

A recent poker tournament supports this interpretation. Poker is far more complex than coin flips. Each player sees only part of the evidence and the odds vary with the strategies chosen by others. This places a premium on clever deceit, an area where humans have long excelled. Nevertheless, in 2017 an AI system "Libratus" routed four top human players in a high-stakes poker tournament, thanks largely to its appetite for risk:

> One of the things Libratus does well is bluff…. [T]he system's aggression flummoxed the human pros…. Libratus has also been over-betting frequently, wagering far more to win a hand than is currently up for grabs in the pot. "If you have $200 in the middle and $20,000 in your stack, you can bet that," says Doug Polk, a poker pro who bested a previous AI…. "But humans don't really like that. It feels like you're risking a lot of money to win so little. The computer doesn't have that psychology. It just looks at the best play." (Popper 2017)

In capital markets, a slower-than-optimal pace of change offers opportunities for arbitrage. When price gives evidence of trending, bet that the trend will continue. That is momentum trading. As noted in Chapter 1, the long-standing success of momentum trading baffles orthodox finance. Market sluggishness from fractional Kelly provides a clue. The longer-term the security, the bigger role this seems to play. Insight about tomorrow's likely dividend gives insights about likely changes in price of the rest of the security. For example, if the dividend is higher than expected, future dividend streams will likely be repriced higher too. However, disagreement is rife, and people who disagree never fully agree on what they disagree about and why. This encourages deference to the status quo.

Granted, differing interpretations of the latest observation might not provide sufficiently sustained trends for momentum trading to capitalize on. However, there is another kind of deference that seems ripe for momentum trading. Sometimes research changes our views on long-past events in ways that reshape our views about the future. Even if we are confident that others will come around to similar views, we can't be sure how long it will take, and markets can stay wrong longer than knowers can stay liquid. This encourages us to study relevant market movements for evidence of an inertial trend and to adjust the size and timing of our bets to its momentum.

The flip side of successful momentum trading is that it encourages bets on spurious trends too. Fantasies about long-term development can flourish because confirming or refuting observations take so long to arrive. For example, imagine a security that pays the world's average temperature every day for the next 100 years. People already dispute the strength of the current warming trend, its relation to CO_2 emissions, and the likelihood of acceleration. Rational myopia predicts that the security price will depend far more on relatively recent temperatures than the duration of payments suggest. Whichever way the price trends, champions of the opposing view are bound to scoff at the shortsightedness.

5

Fitness versus Robustness

This chapter likens the evolution of beliefs to biological evolution. Stable environments favor static fitness. Unstable environments favor robust adaptation. On the advice of a dictionary, the title shortens static fitness to fitness ("ability to survive and reproduce in a particular environment") and robust adaptation to robustness ("ability to withstand or overcome adverse conditions"). However, don't let brevity mislead. Adaptation is vital for dynamic fitness, or "antifragility" as Taleb (2012) called it, whereas robustness can suggest imperviousness to change.

In capital markets, static fitness is best served by matching the mean belief E to the actual risk θ and trimming the variance V around it. Robust adaptation is best served by a V that surges quickly when θ shifts so that E can catch up sooner. V can best surge if skewness surges in the direction that θ moves. Conversely, V should retreat when E catches up to θ, which is facilitated by a skewness that retreats too. In general, the cumulant hierarchy helps tune the pace of learning to slash large errors quickly or grind down small errors slowly.

The tension between static fitness and robust adaptation posed a huge challenge for Darwinism. If evolution favors survival of the fittest, how does it cope with fitness in one environment becoming ill-suited to a new one? Does it modify a dominant trait or species or simply replace it? Is adaptability itself an inheritable trait? if so, might evolution select for it?

Charles Darwin recognized the challenge but couldn't resolve it. He thought inheritance occurred mostly through pangenesis, whereby sperm captures all the father's traits, eggs capture all the mother's traits, and fertilization blends the traits together. Under pangenesis, reproduction generates averages of averages. This tends to smother variation long before natural selection can act.[4] Darwin inferred that reproductive systems must be inherently fragile and prone to random mutation.

In contrast, Mendelian genetics seemed to defy natural selection. Each organism carries many pairs of trait-determining genes, with one gene in each pair inherited from the mother and the other from the father. While one gene typically dominates the other in expression, the two genes are equally likely to get transmitted to offspring. Thus, sexual reproduction preserves variation from one generation to the next but appears to undermine the survival of the fittest.

The split was resolved by Ronald Fisher, arguably the greatest of Darwin's successors (Dawkins 1986, Edwards 2011) and also a giant in statistics—the maximum likelihood method was just one of his many innovations. He called the resolution the "fundamental theorem of natural selection" and stated it as follows:

> The rate of increase in fitness of any organism at any time is equal to its genetic variance in fitness at the time. (Fisher 1930)

To understand its gist, consider two competing genes in proportions p and $1-p$. Suppose that the first gene outcompetes the second at rate α. It is readily shown that

$$\frac{dp}{dt} = \alpha p(1-p), \tag{5.1}$$

which indeed equates the rate of change to variance. Figure 5.1 charts it as a function of fitness. To be clear, this depicts a statistical regularity rather than a deterministic outcome. While Fisher understood this well, other evolutionary biologists of that era were much less attuned to statistics and slower to appreciate the implied tradeoff between static and dynamic fitness.

Figure 5.1: Simplified version of Fisher's Fundamental Theorem

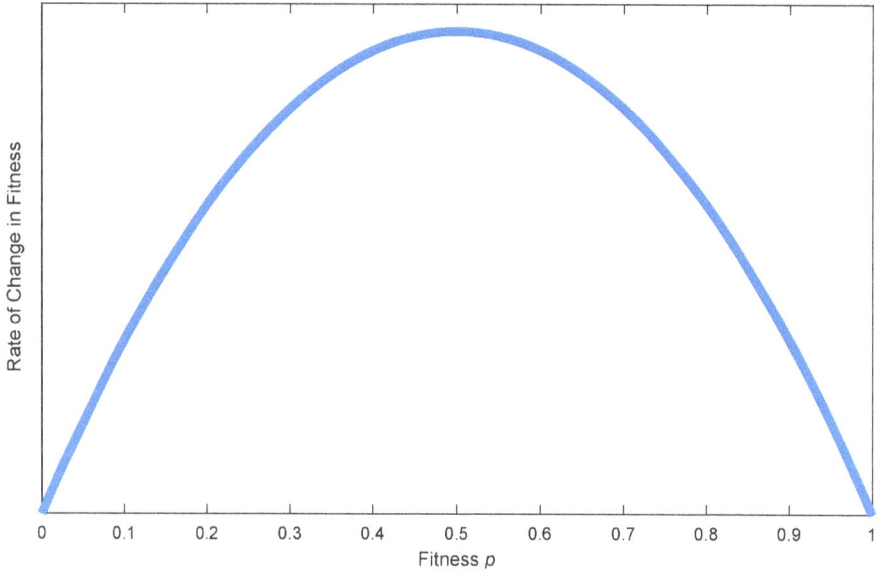

Pressures for static fitness tend to purge the worst genes, which if too successful reduces variance and slows adaptation. Adaptation is fastest where neither gene clearly predominates. At first glance, this is just a transitory problem. Fitness keeps increasing until $p=1$, where only the fittest survive and no more variance is needed. However, as Fisher emphasized, the environment constantly changes. The very success of a species in exploiting its environment tends to tilt the environment against it, e.g., by overgrazing pasture or hunting prey to extinction. Variance is key to helping species back out of evolutionary dead ends.

Fisher's Fundamental Theorem turns out to have a close connection to the Core Learning equation. To establish it, let's recast the efficiency edge α as a fixed difference A times a random stress $x - E$ relative to the current mean. Let's also divide by the variance experienced in a short time interval Δt, since that controls for the units of measurement. The combination transforms (5.1) into

$$\Delta p \approx p(1-p)A\frac{x-E}{\sigma_E^2} + \langle\text{mutation-induced shift in } p\rangle,$$

which is what (3.1) boils down to for a two-belief system with $A \equiv \theta_1 - \theta_0$. Genetic mutation provides the migration that completes the analogy.

Fisher's focus on variance was a much-needed corrective to blended inheritance. It helped end the war between Darwinians and Mendelians. It rebuked some primitive views on genetic superiority. Variance allows gradual changes to make qualitative leaps over time without demanding occasional miracles, such as a fully functioning eye popping out of nowhere.

However, Fisher pressed the argument too far. He dismissed small, relatively isolated populations as evolutionary backwaters, too prone to uniformity or extinction. While these risks are real, smaller populations are more likely through random genetic drift to cultivate new features. This can make them hothouses of genetic innovation. Sewall Wright (1932), another great evolutionary biologist, emphasized their usefulness, prompting bitter disputes between him and Fisher.

What Fisher missed is that variance can be even more useful when it's flexible. Stress tends to skew the gene pool, which can help ratchet up variance to better respond to stress. When stress recedes, skewness fades. The combination provides yet another kind of evolutionary dynamism. It lets natural selection foster static perfection when the environment is stable and substantive change when the environment shifts. When Richard Neher and Boris Shraiman (2011) generalized Fisher's theorem to the full cumulant hierarchy, they found support for some of Wright's core insights.

Variance and higher cumulants are just as important to learning in capital markets as they are to biological evolution. Variance explains excess price volatility while higher cumulants explain its ebb and flow. Unfortunately, standard finance models confine variance to a narrow realm. Everyone either agrees on risks and rewards from the outset or converges to agreement as evidence rolls in. This is the finance analogue of pangenesis. To revive variance, beliefs must occasionally diverge, as they will when trends change without warning. Given enough evidence of regime shift, Bayes' Law can generate abundant heterogeneity out of tiny slivers of doubt.

Figure 5.2 provides a stylized illustration. Like in Chapter 3, it depicts mean beliefs E for a coin that switches to $\theta = 0.9$ from an initial $\theta = 0.5$, with observations numbered after the switch. However, observers are no longer confined to two beliefs and the pattern after the switch need not be regular. Instead, 100 observers view the same sample path, including 20,000 observations before the switch, and update their beliefs rationally. No two observers agree on both the probability of switch and the distribution θ is resampled from. Their differences span ten reset rates from 10^{-2} to 10^{-11} and ten distributions from beta $(1,1)$ to beta $(10,10)$.

While mean beliefs are close at the beginning and relatively close at the end, they diverge widely during the turbulent transition. How should we form a consensus predictor? Figure 5.3 compares four methods. The first sticks with the best predictor overall. Since stability paid off before the switch, it is the slowest to react after the switch and has the highest mean squared error. The second method selects the median prediction, which adjusts much faster. The third method selects the simple average, which starts to shift even sooner but never moves fast and drags out adjustment

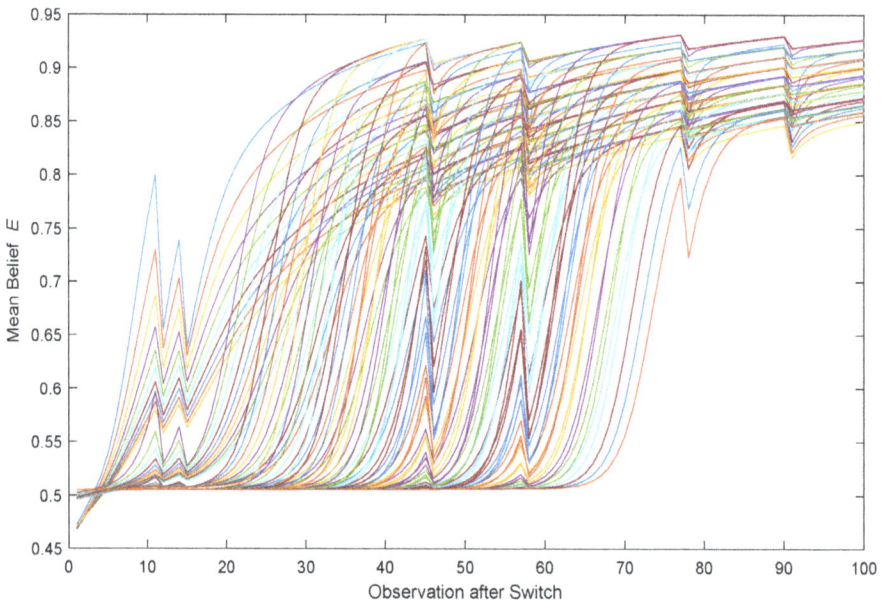

Figure 5.2: Mean beliefs after switch from $\theta = 0.5$ to $\theta = 0.9$ with disagreements over reset process

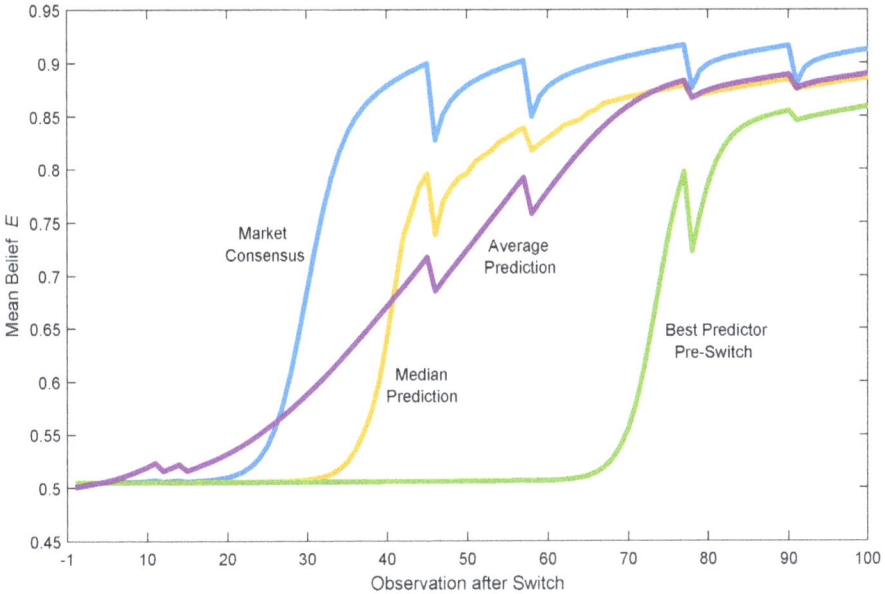

Figure 5.3: Comparison of market, median, mean and best pre-switch predictors for Figure 5.2

too long. The fourth method relies on a market consensus that allows a modicum of doubt. While it is by far the most skittish method it tracks the underlying risk best. Here are larger lessons:

- Don't stick forever to tried-and-true, as it might cease to be true.
- Don't ignore outliers, as they might give early warnings.
- Don't give all views equal credibility, as they might not deserve it.
- Markets help credibility dynamically adjust.

Market mechanisms also encourage collaborative learning. Suppose a modeler has no idea how the market E was generated but discovers a variable z and interpretation $\phi(E, z)$ that predicts θ better than E does. From an evolutionary perspective we can regard ϕ as an adaptation of E that draws on a new resource z. Market mechanisms make it easy to bet profitably on this, which in turn encourages researchers to look for candidate z and ϕ.

One popular candidate, logistic regression, typically treats the log odds y as polynomial in z, converts it into a probability estimate $\phi_L(y)$, and adjusts the parameters through iteration against the various outcomes x. Efficient adjustments have mirror images in markets of rational gamblers, where each gambler knows only a single piece of a larger puzzle. Since logistic regression has parallels in animal nervous systems and is a core building block of the neural networks used in machine learning, the mathematics likens collaborative learning to the evolution of natural and artificial intelligence.

For a simple example, suppose our evidence z matches θ exactly but we believe that it is linear in y, so we plug $a + bz$ into the logistic ϕ_L and look for the best a and b. Despite the false presumption, the resulting estimator isn't bad overall. The absolute deviation from θ averages less than 0.03, although it widens to 0.07 at the edges. Later, someone finds that adding a cubic term in $z - 0.5$ to y improves the fit. As refinements proceed, the inputs typically get more complex and the causal connections more obscure. The market doesn't care. It just aggregates the credibility-weighted inputs before an observation and reshuffles their credibility after. See Figure 5.4.

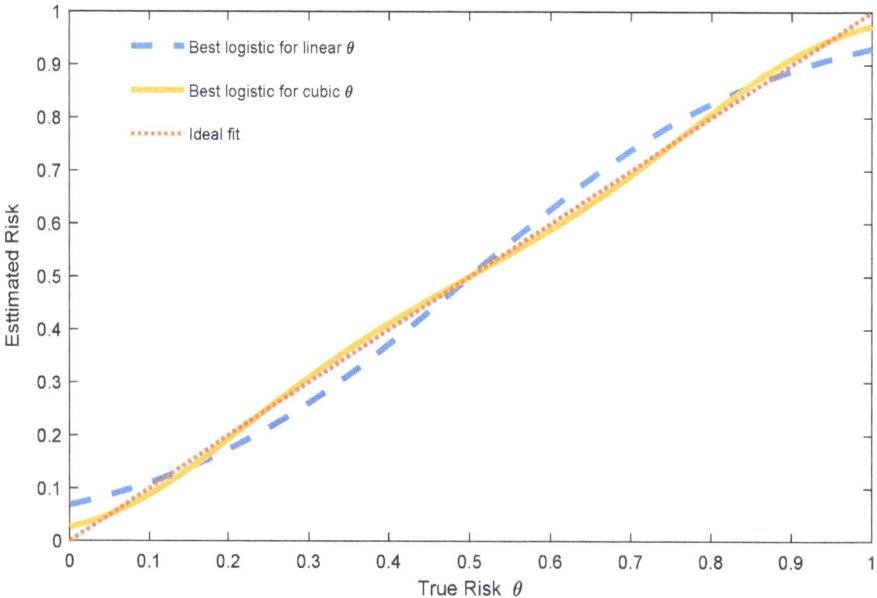

Figure 5.4: Best logistic predictors when log odds falsely presumed linear or cubic in the true risk.

Short Term versus Long Term

The exquisite complexity and order of nature often appears to reflect well-planned, intelligent design. Modern biology counters that natural evolution is blind. Organisms carry vestiges of their origins long after the need for them is gone or fail to develop traits that would be extraordinarily useful. What perfection we observe is the polish of repeated clashes, like rocks ground for eons in a tumbler.

The most glaring imperfection is the natural selection process itself. What evolution ideally selects are species that encompass wide enough variation to adapt to new environments and extend themselves through reproduction. What evolution actually selects are the luckiest individuals with a bias toward static fitness.

Robustness and reproduction have ways to sneak back into favor. Recessive mutations may persist through enough generations that they combine for dominance and prove their evolutionary worth. Altruistic behavior might be passed on through siblings even if the noble altruist dies without progeny, or various self-interested behaviors might join to create the appearances of altruism. Adaptability itself will be selected for if less adaptable species perish.

Yet such concerns are peripheral to the immediate struggle for survival and their evolutionary patches take longer to ripen than the gambits of predators and prey. The best patch is sexual reproduction, which helps preserve genetic variation. It took hold after an estimated billion years of life without it. Why did it take so long? Because asexual reproduction was easier, didn't create an egg-less gender and selected more directly for the genes of the statically fit.[5]

Mother Nature didn't fret over the delay. Mother Nature didn't worry that 100 million years of evolutionary refinement in huge dinosaurs would be wiped out by a single meteor strike. Mother Nature didn't anticipate that small mammals would better survive the cataclysm. It placed no bets that humans would evolve from small mammals or that humans would distinguish themselves from other species by their sociability and ingenuity.

Tension between short-term survival and long-term vigor is evident in capital markets too. Although capital markets reward short-term forecasts well, feedback on long-term forecasts is slow and interim payoffs depend mostly on others' beliefs about future. Keynes (1936) famously likened securities markets to beauty contests in which judges pick the contestant they expect other judges to gauge as most beautiful.

Great seers are rare and focus on what can't help but happen. They step back from current news, look at the big picture and seek to identify ineluctable trends. Their forecasts rest on insight rather than foresight. However, market traders aren't the best audience for either. The wisdom of the crowd reviews a host of opinions, filters the latest evidence, and takes a popularity poll weighted by capital and conviction. Traders naturally focus more on getting slightly ahead of the crowd than on deciding where to lead it.

While the rarity of great long-term forecasts increases their potential worth, this is better tapped through private equity claims than through public markets. The latter provide indirect assistance by allowing early backers to cash out and invest somewhere else. However, venture capitalists can't completely avoid rational myopia either and to some extent they feed it, by gearing their investments toward offerings they think public markets will favor. They too trade on market momentum, only at a longer horizon than most.

In short, capital markets are poor long-term predictors but superb short-term correctors. They don't deserve deep trust. They do deserve respect, as the best mechanism we have for projecting recent trends into the future and adjusting to new evidence.

PART II:
HAROLD'S CASINO

Harold's Casino operates roulette wheels whose odds occasionally change. Tiny doubts catalyze seemingly abrupt reappraisals. Imagine the odds of red or black have long seemed perfectly fair. We might need to see more than a dozen straight reds before we accord even a 5% chance that something is amiss. Yet another dozen straight reds will likely make us 95% confident the odds have changed.

When risk ranges more widely, behavior alternates between calm and turbulent phases. Much rationally turbulent behavior looks irrationally exuberant. Markets of rational gamblers can generate even more turbulence than their participants, even though the whole tends to behave more rationally than its parts. This warns against inferences of irrationality that rely solely on evidence of market froth.

When markets respond sub-optimally to news, they seem more likely to swing too little than too much. The caution reflects a mixture of risk aversion and deference to the wisdom of the crowd. Momentum trading can mitigate that by extending and speeding the aggregate market response. However, no market mechanism effectively compensates for the inherent difficulty of forecasting long-term trends.

6

Benefits of Doubt

To strengthen intuition for the Core Learning equation, Part II explores its operation in the simple games introduced earlier. Their technical name is Bernoulli except that we drop the standard assumption of stable risk. The outcome x_t at time t takes value 1 or 0 with probability θ_t or $1 - \theta_t$ respectively. Let $p_{t|u}$ denote the subjective probability or uncertainty attached to possible values of θ_t given the observations through time u. We will pay special attention to the first two cumulants

$$\text{mean belief} \quad E_{t|t-1} \equiv \left\langle \theta_t \right\rangle_{p_{t|t-1}} \equiv \int \theta p_{t|t-1}(\theta) d\theta$$

$$\text{variance} \quad V_{t|t-1} \equiv \left\langle (\theta_t - E_{t|t-1})^2 \right\rangle_{p_{t|t-1}}$$

After observing x_t, we update beliefs about θ_t using Bayes' rule (2.1) to obtain:

$$p_{t|t}(\theta) = \begin{cases} \dfrac{\theta p_{t|t-1}(\theta)}{E_{t|t-1}} & \text{if } x_t = 1 \\[2em] \dfrac{(1-\theta) p_{t|t-1}(\theta)}{1 - E_{t|t-1}} & \text{if } x_t = 0 \end{cases}.$$

Let $\Delta_b p_{t|t-1} \equiv p_{t|t} - p_{t|t-1}$ denote the backward-looking Bayesian update. Dropping the time subscripts as understood, simple algebra indicates that

$$\Delta_b p(\theta) = p(\theta)(\theta - E)\frac{x - E}{E(1 - E)}. \tag{6.1}$$

Since $\sigma_E^2 = E(1 - E)$ for Bernoulli games, this accords with (3.1). Next we update $p_{t|t}$ to $p_{t+1|t}$ by anticipating net migration in θ over the next period. The simplest adjustment posits a constant probability λ that θ resets to a random draw from a uniform distribution. In that case,

$$p_{t+1|t}(\theta) = p_{t|t}(\theta) + \lambda\left(1 - p_{t|t}(\theta)\right).$$

To trim more notational clutter, let $\Delta_f p \equiv p_{t+1|t} - p_{t|t}$ denote the forward anticipation and write

$$\Delta_f p(\theta) = \lambda\left(1 - p(\theta) - \Delta_b p(\theta)\right). \tag{6.2}$$

The total update $\Delta p = p_{t+1|t} - p_{t|t-1}$ is just the sum of the backward and forward adjustments:

$$\Delta p = \Delta_b p + \Delta_f p \tag{6.3}$$

Now that we have clarified the update mechanisms, let's see how they work at an imaginary casino run by my alter ego Harold. Harold's banner game is roulette played on a gracious wheel. Gracious, because there are no green pockets for the house's customary take, just 18 red pockets alternating with 18 black pockets. Since the pockets are identical apart from color and number and the wheel seems regular, our beliefs cluster around $\theta = 0.5$. However, to minimize biases and wishful thinking, let's track the outcomes before we place any bets.

Fortunately, the casino keeps excellent records, which indicate that of the last 10,050 plays, the ball landed on red 5,050 times. The favored method for using this information amounts to a simple counting game. Imagine an urn with N red balls and N black balls. To this add one ball for every roulette outcome of the same color. The urn now contains $5050 + N$ red balls and $5000 + N$ black balls. If we pick a ball at random, the expected chance of red is $E = \frac{5050 + N}{10050 + 2N}$. The highest N we can assign is infinite, in which case $E = 0.5$ regardless of what occurred at the casino. The lowest N we can assign is 0, in which case $E = 0.5025$. Every other N sends us somewhere in between.

However, a quick review of Harold's records reveals something extraordinary. The last 50 roulette plays landed red, with a chance given independent fair plays of one in a million billion. How should we factor that into our views on θ? Let's ask a succession of bots to analyze the data, to forecast E as best they can, and to justify their methodologies. Our first bot, Classical Purist (CP), sets $E = \theta = 0.5$. Let's see how well it passes a so-called Turing test for semblance to human reason.

KO: Don't you see how rare 50 reds in a row are? Why doesn't that change your estimator?

CP: *The ordering of iid observations doesn't matter. All that matters is the aggregate count of reds and blacks, which is only half a standard deviation away from an even split and well inside any reasonable confidence interval.*

KO: Why do you assume the observations are iid?

CP: *Classical statistics. That's how we reconcile common risk with varying outcomes.*

KO: Are you sure the risk is common? What if something went wrong with the wheel?

CP: *Harold is supposed to take care of that. It's his casino.*

KO: He's devious. Why would I pose this thought experiment if he weren't?

CP: *I don't understand.*

KO: If you were human, I think you would. Human brains evolved mainly in competition to deceive others and keep from being deceived.

CP: *Are you suggesting I can't detect falsification? That's false. Given a long enough sequence of reds, I too would conclude that θ exceeds 0.5.*

KO: How many more straight reds would you need to be 99.9% confident in that conclusion?

CP: *About 250. Granted, that's 300 straight reds in all, with a stand-alone probability under fairness of one in* 10^{90}.

Turing test failure. Our second bot, Bayesian Purist (BP), starts with equal likelihoods for every θ and updates them every flip using (6.1).

BP: *Actually, I use a far simpler shortcut. Each* $p(\theta)$ *stays proportional to a power of* θ *times a power of* $1-\theta$, *so beliefs are beta-distributed with a mean E that is easy to compute.*

KO: How easy?

BP: *Just set* $N = 1$ *in your simple counting game.*

KO: Don't the order of reds and blacks matter?

BP: *Not with* θ *fixed. There I agree with CP. However, unlike CP, I don't cling to* $E = 0.5$. *My current forecast is 0.5025 and I will keep boosting it as long as reds keep appearing.*

KO: Given a long sequence of straight reds, will you get 99.9% confident that $\theta > 0.5$ sooner than CP will?

BP: *Not really.*

Another Turing test failure. By assuming that risk is fixed, both bots rule out the only plausible interpretation of the evidence. During the streak of 50 straight reds, θ must have averaged much higher than 0.5. Even a θ as low as 0.8 beggars belief, since the odds against it generating 50 straight reds are 70,000 to 1. There must have been a huge jump.

Granted, we can't expect to detect the jump when it occurs. The first few straight reds always look like noise. Only later do we recognize a break in trend and work backward to its likely start. There's nothing wrong with doing so; we'd be foolish never to reinterpret history. However, it is inconsistent to look for past breaks without acknowledging that tomorrow might mark something new. At best, we anticipate tomorrow's relevant probabilities. At worst, we doubt our current view is right without claiming to understand what or why.

Tiny Doubts

Our third bot, Tiny Doubter (TD), combines Bayes' rule (6.1) with the jitter of (6.2). Feel free to ignore the precise formulas. What matters most are these four characteristics:

- **It is simple.** Bayes' rule multiplies $p(\theta)$ in proportion to the conditional probability $h(x/\theta)$ of the outcome. Doubt nudges $p(\theta)$ a small fraction λ toward equal weights.
- **It is tedious.** Simple approximations rarely work for long. At every observation we need to update beliefs over a host of θ.
- **It is volatile.** Forecasts bounce around and are highly sensitive to small changes in parameters or observations.
- **It is robust.** Overall it tracks risk remarkably well under a wide variety of scenarios and is particularly good at correcting errors.

TD is best appreciated through simulations. Suppose our review of Harold's records uncovers no gross unfairness in the first 10,000 observations. For example, one sequence of 15-20 straight reds or straight blacks shouldn't alarm, as that is nearly 25% likely given 10,000 fair plays. Let us stylize the records as 5,000 reds randomly interspersed with 5,000 blacks, followed by 50 straight reds at times $t=1,\ldots,50$. I will call this Harold's Surprise. If each red pays 1 and each black pays 0, the fair price on the next outcome x is the mean belief E. Technically, E is the most a risk-neutral trader would pay for x. It is not a true market price since different bots will often disagree about fair price and we haven't yet explored how markets form a consensus.

When the doubt $\lambda=0$, TD is the same as BP. Over the last fifty reds its E rises a measly 0.0025. Yet when $\lambda=10^{-6}$, just one part in a million, the response is much more energetic. Figure 6.1 displays the post-Surprise paths for E in 200 different simulations. Let's ask TD to explain.

TD: *Which one of us should explain? Every curve depicts the path of E for a random shuffle of 5,000 reds with 5,000 blacks. Harold's Surprise doesn't give enough info to narrow down to one. Sometimes we disagree a lot.*

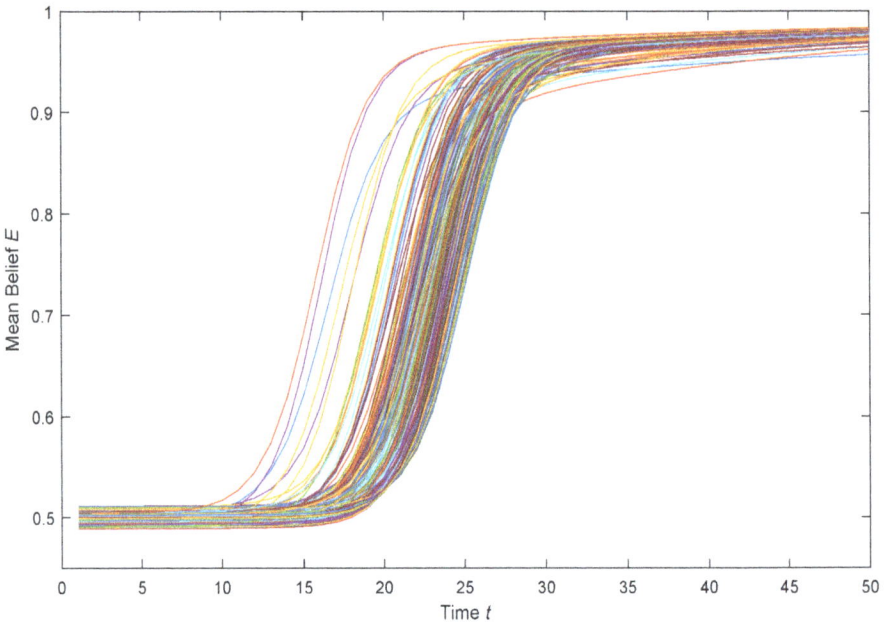

Figure 6.1: Mean beliefs during Harold's Surprise when $\lambda = 10^{-6}$

KO: Yet nearly all of you evolve on a pronounced S-shaped path. Why?

TD: *It takes time to shake our confidence in near-even odds. Once we strongly suspect that θ has changed but aren't sure what it changed to, we get highly responsive to new evidence. As we home in on the new θ, more observations confirm what we already believe and adjustments slow.*

KO: Why do the shapes steepen at such different times?

TD: *The main difference is how many straight reds we observe just before the last 50. Roughly half of us observe none, a quarter observe one, an eighth observe two, and so on. How many reds we saw just before the last black matters too, how many before the next-to-last black and so on.*

KO: Despite the differences in priors, most of the transitions look similar once they start.

TD: *Very. Once E exceeds 0.52, we can be 95% confident that the next 8 or 9 straight reds will send it over 0.9.*

KO: Even the acceleration and deceleration tends to fit a common pattern, as Figure 6.2 shows. How much does this depend on your particular λ?

TD: *Doubling or halving λ shifts the S-curve by approximately one period but keeps the shape basically the same.*

KO: Which λ do you think is best?

TD: *Higher λ is better for responding to Harold's Surprise but induces more jitter when θ is stable. The relative benefits depend on context. The only thing I'm sure of is that $\lambda = 0$ is dangerous, because it isn't robust to shocks.*

Well said. The Tiny Doubters have passed basic Turing tests. As a bonus, they also simulate swings between complacency and overreaction that behaviorists attribute to humans. Specifically, TDs exhibit:

- extended periods when new evidence apparently fails to register;
- large variations in the onset of adjustment;
- sharp acceleration after adjustment begins in earnest.

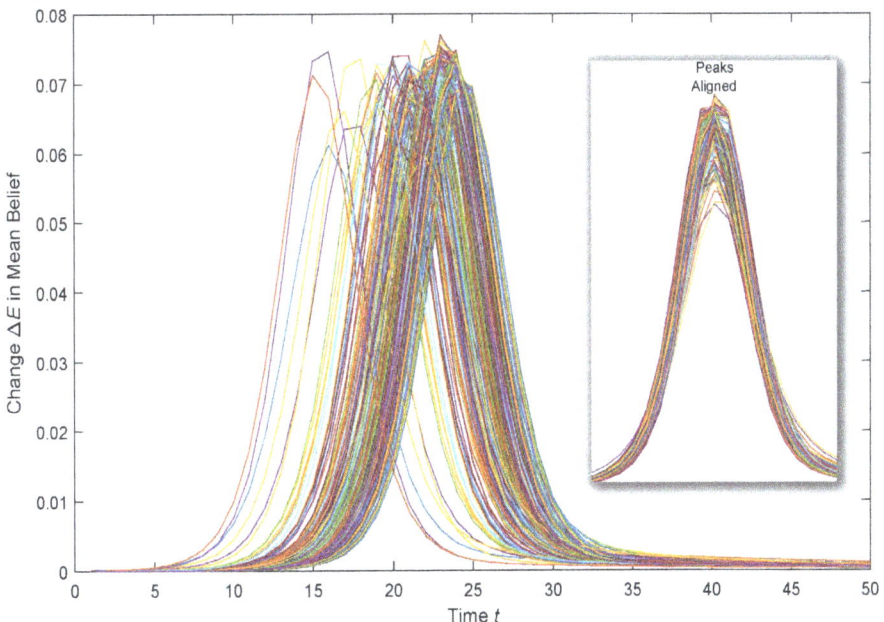

Figure 6.2: One-period changes in mean beliefs for Figure 6.1

To better appreciate the rational drivers, imagine that TDs are torn between two hypotheses: $\theta \approx 0.5$ and $\theta \approx 1$. By Bayes' rule, the odds of the latter approximately double with every red. Initially, $p(1)$ might be so tiny that the doubling is barely noticed. Yet once the odds reach 0.2, eight consecutive reds will thrust them to 5. We thus have a simple explanation of S-curve growth. We also see why λ matters: it sets a lower bound for $p(1)$. While no two hypotheses can fully explain TD responses to Harold's Surprise, the likelihoods always get highly bifurcated during the surge. See Figure 6.3 for representative examples when $\lambda = 10^{-6}$.[1]

Suppose that, contrary to Harold's records; the 50 straight reds were interrupted by blacks at $t = 17$ and $t = 34$. Figure 6.4 charts TD responses for $\lambda = \{10^{-3}, 10^{-6}, 10^{-9}\}$. Sometimes the interruption has a huge impact. Sometimes it doesn't. The contrasts seem to beg behaviorist explication. Yet all TD responses are rational by construction.

For a useful categorization, let's focus on the magnitude of changes ΔE, like those depicted in Figure 6.2. Recall from Chapter 3 the definitions

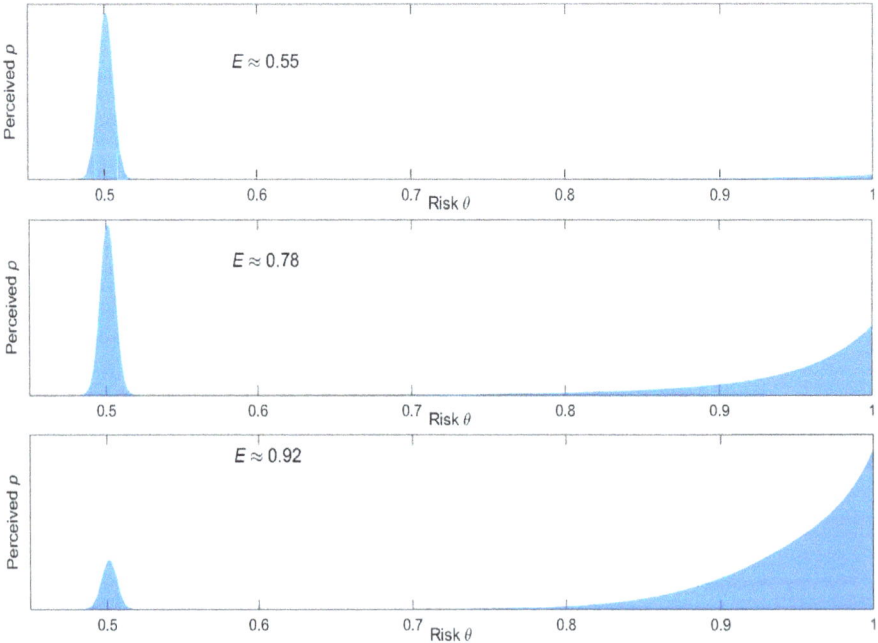

Figure 6.3: Typical beliefs during Harold's Surprise for Figure 6.1

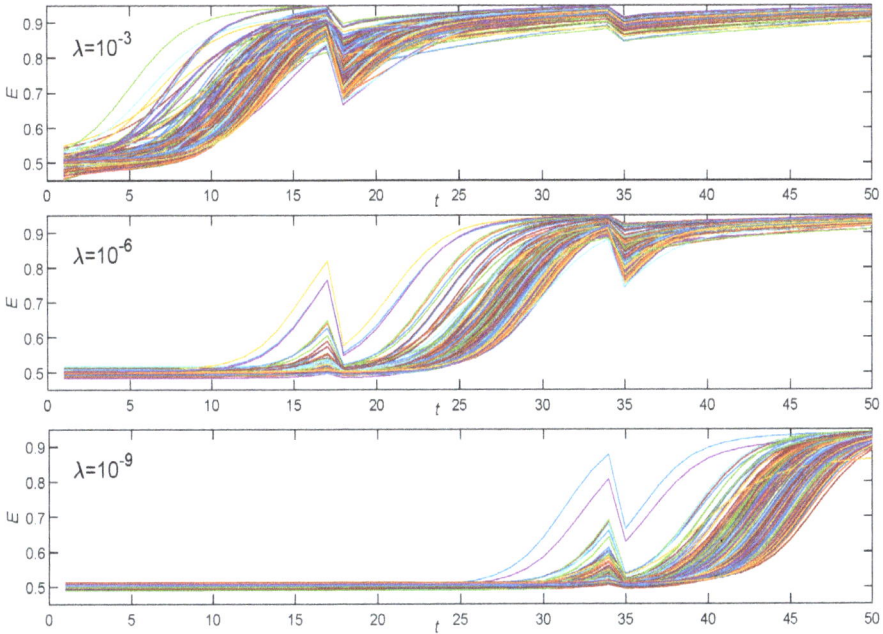

Figure 6.4: E during Harold's Surprise with $x_{17} = x_{34} = 0$ and three different λ

of learning rate η, effective sample size T_{eff}, and variance V of beliefs. Combining them with (3.2) and (6.1) indicates that

$$\eta \equiv \frac{1}{T_{\text{eff}}} \equiv \frac{\Delta E}{x - E} \approx \frac{V}{E(1-E)} \tag{6.4}$$

For an alternative interpretation of η, suppose we treat the true risk θ as a random variable and estimate $\theta - E$ using the proxy $\beta(x - E)$. The mean squared error is minimized when $\beta \approx \eta$. The main implication of (6.4) is that fast learning requires relatively high variance of beliefs.

While η and T_{eff} convey the same information, T_{eff} is more intuitively appealing, since it indicates the perceived number of iid-equivalent observations. For Bayesian Purists, T_{eff} increases by one per observation, so learning always slows. For TDs, T_{eff} plummets when enough straight reds occur and usually bottoms out below four. Figure 6.5(a) plots T_{eff} for Figure 6.1 on a logarithmic scale while Figure 6.5(b) does the same for the middle chart of Figure 6.4.

59

Figure 6.5: Scatter plots of T_{eff} for Harold's Surprise when $\lambda = 10^{-6}$

(a) No black in last 50 plays

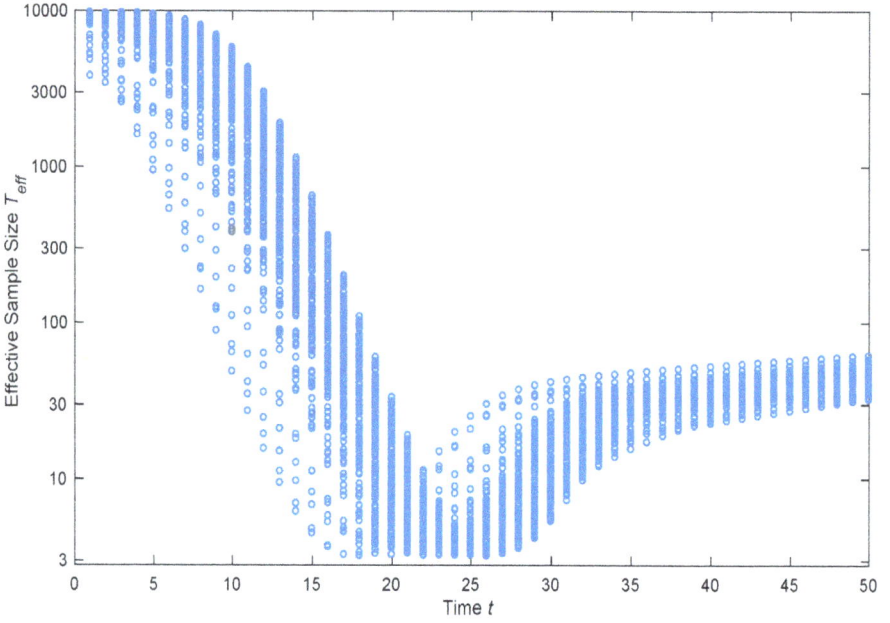

(b) Black at $t = 17$ and $t = 34$

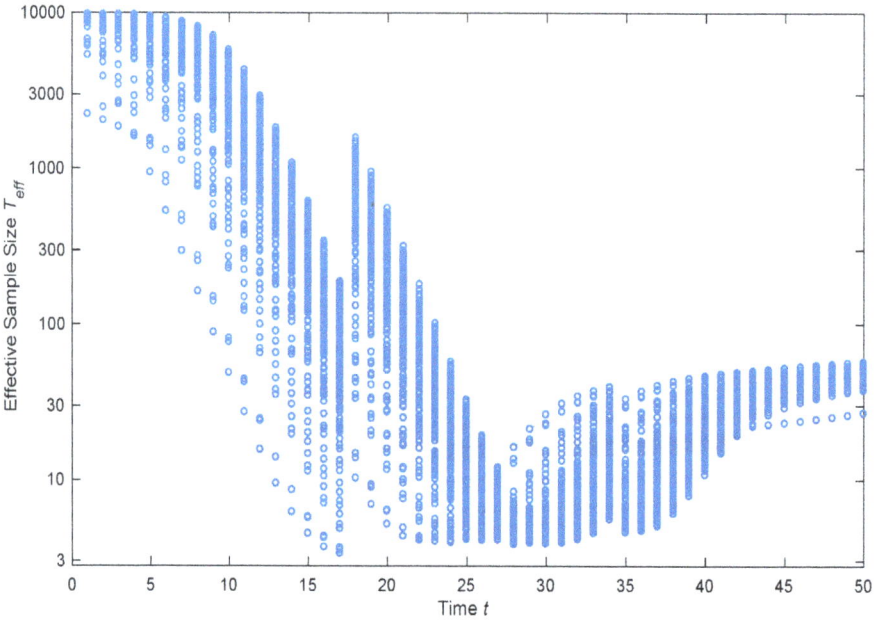

7

Recurrent Shocks

Harold's Surprise shows how rational observers change their minds. They dismiss initial warnings as outliers. Persistent warnings eventually jolt them into realizing that the trend has likely changed. However, the jump from $\theta \approx 0.5$ to $\theta \approx 1$ is too isolated and extreme to reveal long-term dynamics and once people figure out the Surprise it bores.

To entice repeat business, Harold lets graduates play games that vindicate their doubts. After each outcome, θ resets with small constant probability Λ to a new value chosen randomly from a uniform distribution. Otherwise θ stays the same. We first viewed this model in Chapter 4, where $\Lambda = 0.001$ and most base-level learners were EMAs. Here I will set $\Lambda = 0.005$ and focus mainly on TDs.

Let's start with Perfect Learner PL, who sets doubt $\lambda = \Lambda$ in (6.2). Figure 7.1 displays θ and E for 2000 consecutive plays. When θ jumps to extreme values, E responds much faster than in Figure 6.1. That's because PL is primed here to anticipate more frequent shifts: once every 200 plays on average instead of once every million plays.

To slow reactions, we need to model Tinier Doubters. Figure 7.2 displays mean beliefs for λ ranging from 10^{-5} to 10^{-8}. Their doubts understate the average frequency of change by factors of 500 to half-a-million. Even at these levels, the Tinier Doubters respond to changes in risk far faster than Bayesian Purists would. Still, their E lags far behind θ.

Figure 7.1: Perfect learning for 2,000 plays given probability $\Lambda = 0.005$ of reset from a uniform distribution

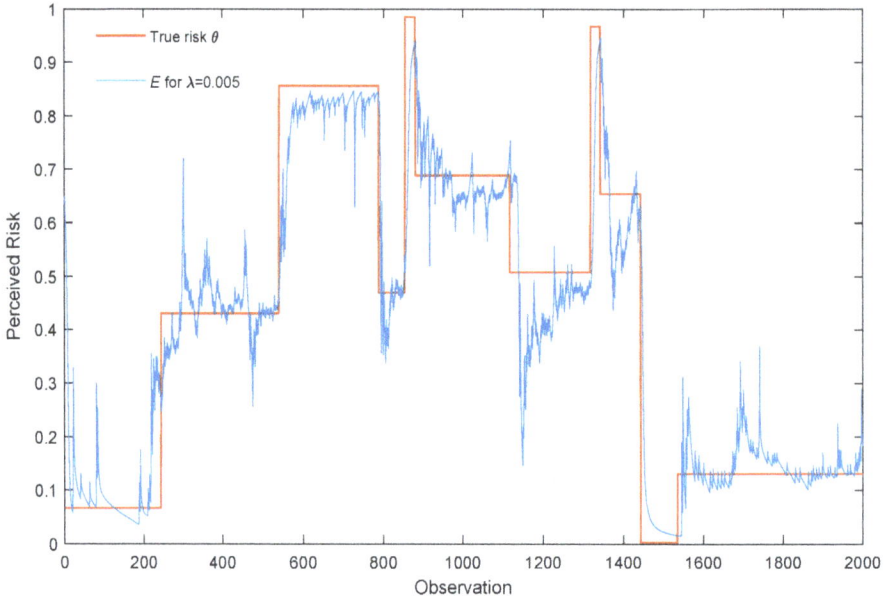

Figure 7.2: Mean beliefs for Figure 7.1 for Tinier Doubters

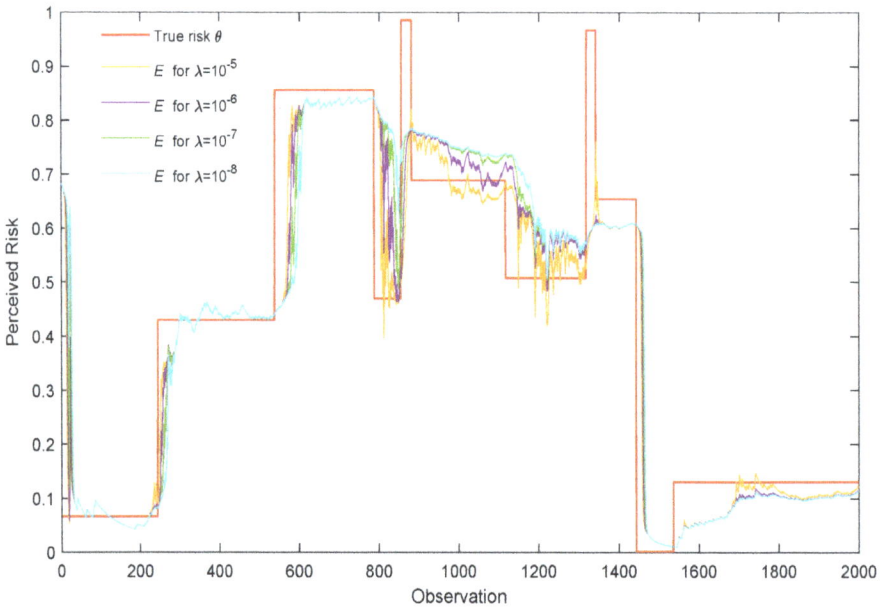

Of course, it's possible to take too much alarm. This happens when λ gets too high. Figure 7.3 charts responses for a Big Doubter with $\lambda = 0.05$, ten times Λ. Even when risk is stable, Big Doubter's estimates rattle over a range at least 0.2 wide. Also, they're noticeably biased toward the center and never get within 0.05 of an edge. On the bright side, the root mean squared error or RMSE is no worse for a Big Doubter with $\lambda = 0.05$ than for the best Tinier Doubter from Figure 7.2 with $\lambda = 10^{-5}$. Both are roughly 30% higher than optimal.

Figure 7.4 displays mean beliefs for $\lambda = 0.01$ or $\lambda = 0.001$. The RMSE for $\lambda = 0.001$ is only 6% higher than the RMSE for PL. The bigger doubt of $\lambda = 0.01$ generates mean beliers that look far more skittish and irrational. However, the chart is misleading. It lets us see the future of every shift, which makes false starts look silly. Even PL in Figure 7.1 looks chronically imbalanced. In fact, the RMSE is less for $\lambda = 0.01$ than for $\lambda = 0.001$. This indicates that false starts matter less than the ability to correct errors quickly. This is true more generally and help explains why capital markets perform as well as they do.

How should we choose an appropriate λ? Suppose we could match the empirical frequency of shifts in θ. Over the course of 2000 plays with $\Lambda = 0.005$, there is only a 0.1% chance of fewer than two shifts or more than 20 shifts. That would keep us safely in the range of Figure 7.4, with only minor deviations from optimality. However, since we can't directly observe θ or count its shifts, the best we can do is to formulate various hypotheses about λ and test them against the evidence.

To see how this works, let's create 12 TDs with $\lambda_i = 10^{-i}$ for $i = 1,\ldots,12$ over a fine grid of θ spaced 0.0002 apart, with initially equal likelihoods. For each of 100,000 plays with $\Lambda = 0.005$, let's update their credibility using (6.1)-(6.3), with an anticipated $\lambda_{TD} = 10^{-6}$ chance in (6.4) that Λ resets randomly to a newly chosen λ_i.

Note that the setup is not well chosen. At every reset, over half the weight goes to TDs that track changes far too slowly, while 8% goes to a TD that jitters far too much. If all TDs maintain equal weights, the overall fit will be poor. Fortunately, our market-like mechanisms downgrade poor TDs quickly.

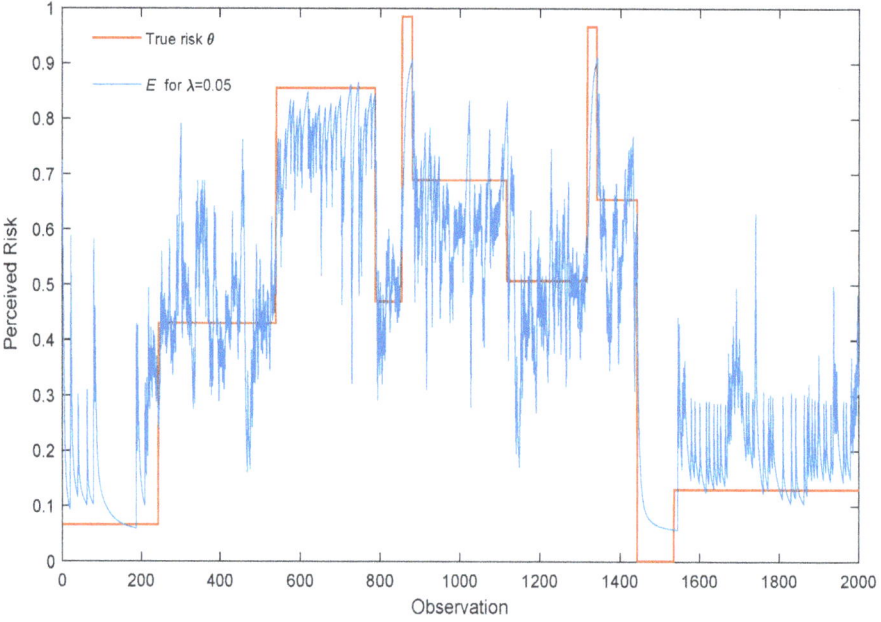

Figure 7.3: Mean beliefs for Figure 7.1 for a Big Doubter with $\lambda = 0.05$

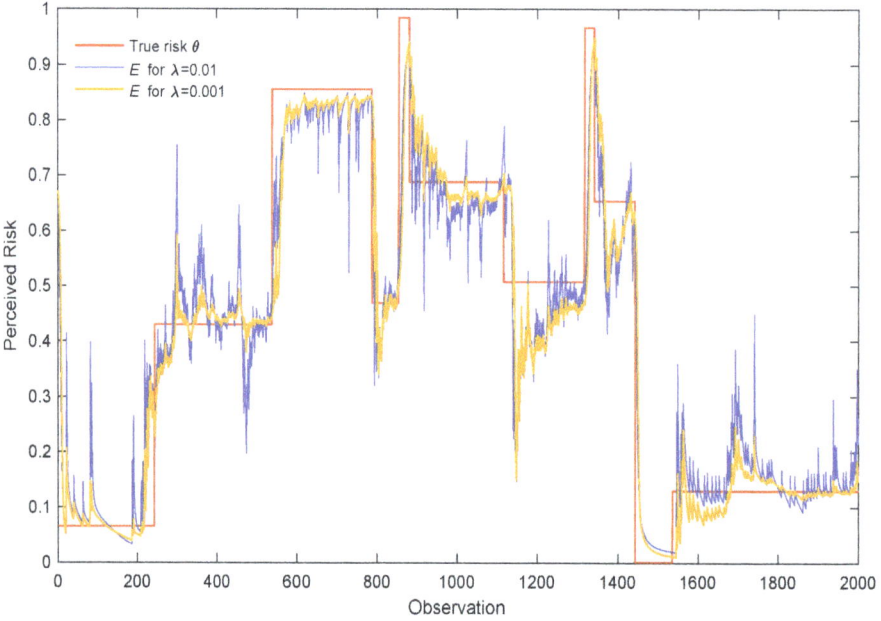

Figure 7.4: Mean beliefs for Figure 7.1 for $\lambda = 0.01$ and $\lambda = 0.001$

The aggregate predictor amounts to a composite, two-layered learning machine with TDs over θ on the bottom and a TD over TDs on top. I will call it Tiny Doubter Two-Layer or TD2L. Note that TD2L makes three bad judgments. It wrongly assumes that one of the bottom TDs correctly identifies Λ, its initial consensus greatly understates Λ, and its updates anticipate changes in Λ that don't occur. Nevertheless, TD2L converges quickly on $\lambda = 0.01$, the best TD of the bunch, with only 2% higher RMSE than for PL. If we redo TD2L with 24 TDs on the lower layer with $\lambda = 10^{-i/2}$ for $i = 1, \ldots, 24$, its RMSE is only 0.2% higher than for PL.

TD2L demonstrates that uncertainty about Λ need not cripple our estimation of θ. However, it can be thrown severely off balance if Λ suddenly changes by a factor of 10 or more. For extra robustness we can let λ_{TD} range over a spectrum of values, each generating a slightly different TD2L, and let a Tiny Doubter aggregate over the TD2L to create a three-layer learner I will call TD3L. To test it, imagine that after 100,000 plays, Λ suddenly drops to 0.0001, just 2% of its previous value. Let's create a TD3L in which λ in the lower layer and λ_{TD} in the second TD layer both range over $\{10^{-1}, \ldots, 10^{-12}\}$. Figure 7.5 displays mean beliefs for both TD3L and PL over the next 100,000 plays. TD3L completes most of its adjustment within 2,000 plays. After 10,000 plays it is hard to distinguish TD3L from PL.

One disadvantage of layered learning is that it obscures the base-level drivers. On the bright side, it makes bots sound more human.

KO: How did your forecasting team manage such excellent tracking?

TD3L: *Through extensive study, consultation and debate.*

KO: Why the occasional overreactions and false starts?

TD3L: *Nobody's perfect. We do try to correct errors quickly.*

KO: What triggers the lapses?

TD3L: *I'm not sure. Seems to depend on context.*

KO: Might your internal controls be flawed or your experts prone to panic?

TD3L: *Perhaps.*

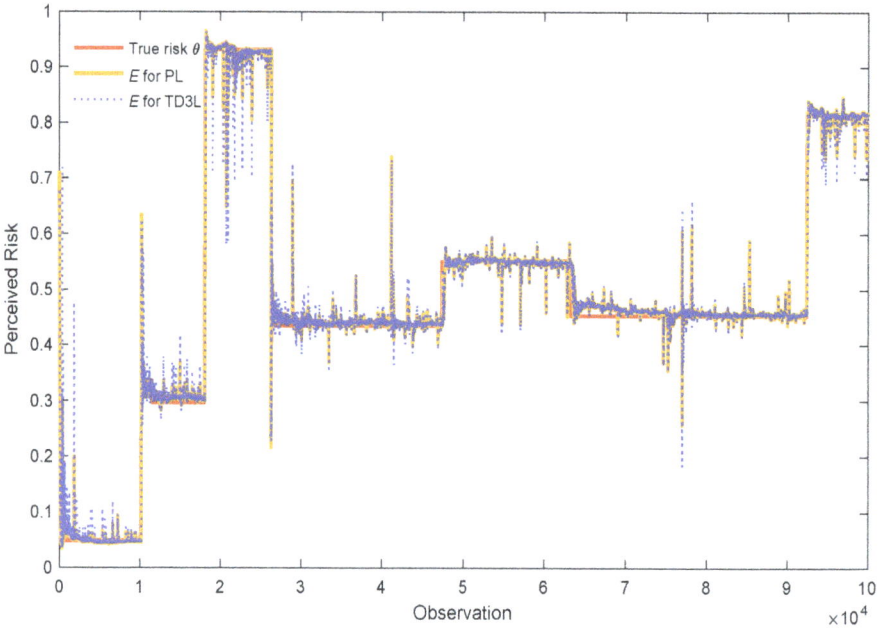

Figure 7.5: Mean beliefs for PL and TD3L for 100,000 plays with $\Lambda = 0.0001$, when TD3L is trained for 100,000 plays prior with $\Lambda = 0.005$

Measures of Turbulence

One measure of turbulence is the compression of effective sample size. Figure 7.6 plots T_{eff} on a log scale for Figure 7.1. Nearly all values lie well below the dotted lines, which indicate the true duration of current risk. In long simulations, the median T_{eff} is 40, just over 20% of mean duration. About 8% of T_{eff} are less than 10. The gyrations they induce can easily be misread as panic.

When shifts in θ are rarer, T_{eff} is larger but more compressed relative to true duration. Figure 7.7 displays T_{eff} for Figure 7.5. The median T_{eff} is 1450, which is 16% of duration. Still, T_{eff} occasionally plunges to extreme lows, with 7% less than 200 and 1% less than 50.

Figure 7.8 orders the learning rates η for Figures 7.5 and 7.7 from fastest to slowest and displays them by their cumulative share of total learning rates. If we measure the influence of an observation by the η associated with it, the most influential 1% of observations account for a quarter of total learning while the most influential 5% account for half.

Figure 7.8 also displays the corresponding shares if duration were known. The most influential 1% of observations, which occur just after θ shifts, would account for half of total learning. Hence, even though most learning occurs in turbulent phases, it would be even more concentrated given better information about changes in risk. The skittishness of Tiny Doubters is largely just catch-up from what they missed before.

Recalling (6.4), the prime driver of fast learning is high variance of beliefs. However, we need more than that to cause turbulence. Imagine two quarreling puppets Punch and Judy, with Punch's initial convictions directly proportional to θ and Judy's directly proportional to $1-\theta$. Their mean beliefs are initially $\frac{2}{3}$ and $\frac{1}{3}$ respectively, so their disagreement is high, and their initial learning rates are high as well. Nevertheless, if both Punch and Judy update beliefs rationally without error or doubt, their beliefs converge smoothly. No matter what they observe, after t plays their disagreement shrinks to $1/(3+t)$ and their learning rate shrinks to $1/(4+t)$. That is more choreography than chaos.

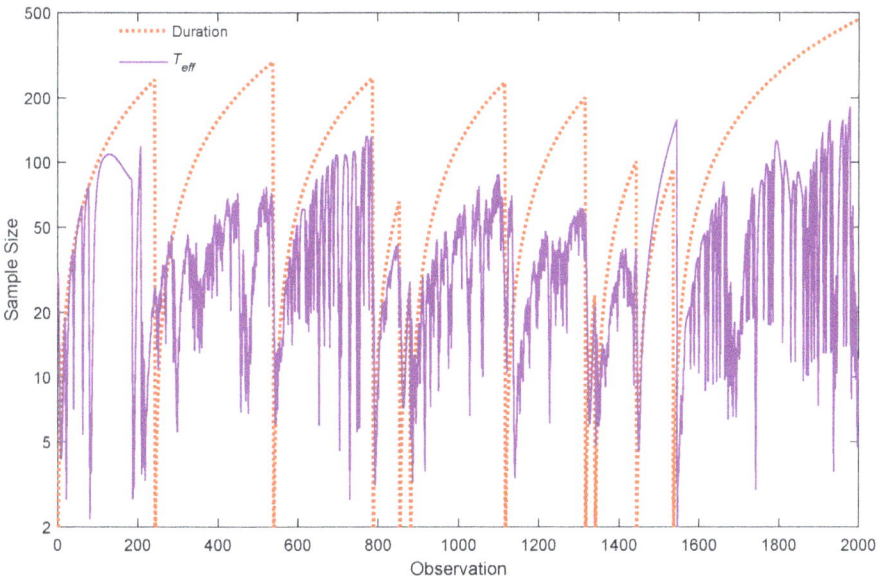

Figure 7.6: Effective sample size for Figure 7.1

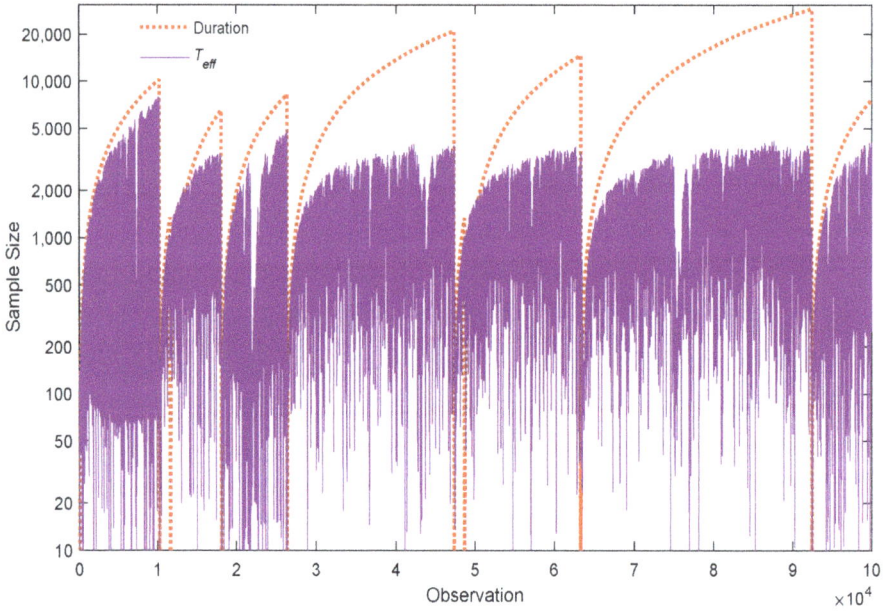

Figure 7.7: Effective sample size for Figure 7.5

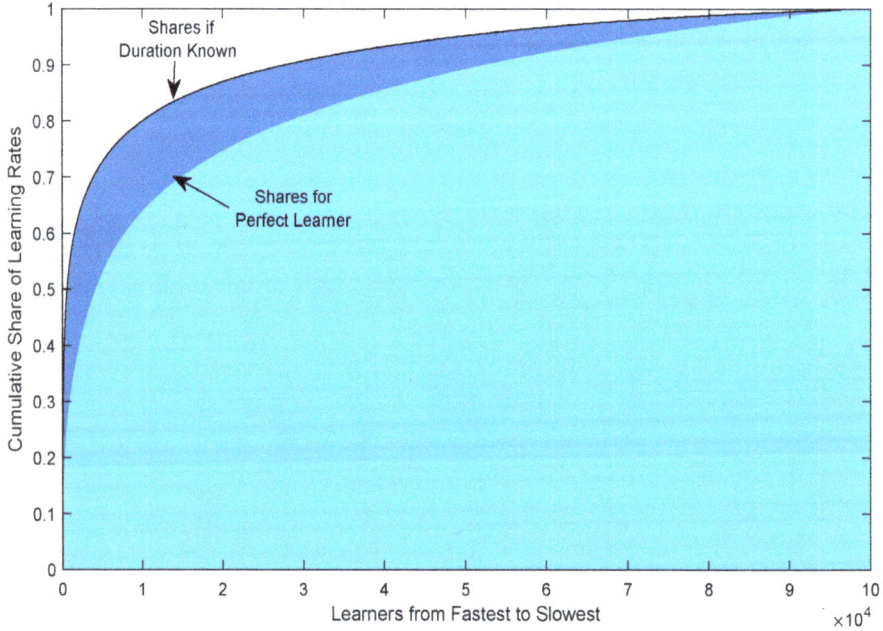

Figure 7.8: Cumulative shares of total learning rates for Figure 7.5

What turbulence requires is a kind of bifurcation of beliefs, like in Figure 6.3. One hump adheres to the risk that previously prevailed and views recent deviations as meaningless bad luck. The other hump defects to the new mean evidenced in the deviations. Turbulence arises when the middle ground between them is relatively bare: either extreme is more plausible than the average.

A similar situation arises when people disagree so sharply that they don't see the point of compromising. If the new evidence persists, we will applaud the prescience of early converts. If it doesn't, we will applaud the wise caution of the crowd. In the interim, perfectly reasonable people will lament others' apparent foolishness.

Of course, this doesn't prove that all disagreement is reasonable or that all market turbulence is rational. The best argument for rationality is natural selection: over time, more rational learners ought to win market share from less rational learners. While all humans occasionally drink from a fountain of wrong-headedness, it is hard to believe that they all drink to the same degree or that they never find reasonable antidotes. From a learning perspective, turbulence is just the froth from major reappraisal and correction.

Look again at Figure 7.1. The graph of E is far more jittery than the graph of θ. If we extend this over a million plays and trace all the changes with thread, the E thread will be 11 times longer than the θ thread. That is a core difference between learning and knowledge. A quarter of the length of the E thread will come from the turbulent top 4% of changes. When we don't understand the causes, it is tempting to ascribe intense bursts to temporary insanity. The false starts in Figure 7.1 seem to flag blatant disregard of the clearly prevailing trends. However, trends are never as clear when they emerge as they appear in hindsight.

The best way to appreciate turbulence is to try to learn without it. Suppose we fix T_{eff} at various constants to create multiple EMAs. Figure 4.4 showed that pure EMAs don't track reset at rate 0.001 well; even the best fit in Figure 4.5 was disappointing. As Figure 7.9 shows, EMAs work no better when $\Lambda = 0.005$. Setting $T_{eff} = 1/\Lambda = 200$ is hopelessly slow-moving.

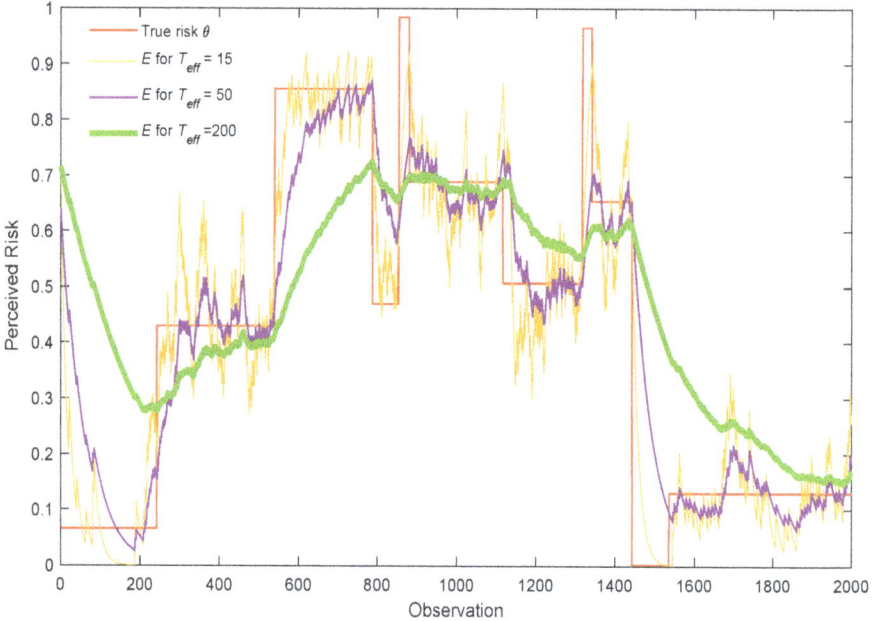

Figure 7.9: Mean beliefs for three EMAs given the observations for Figure 7.1

Setting $T_{eff} = 50$ to match its average value in simulations predicts E worse than a TD with $\lambda = 10^{-6}$. The best single EMA predictor sets $T_{eff} \approx 16$ but predicts E no better than a Tiny Doubter with $\lambda = 10^{-4}$.

Pure EMAs never fit the data well when risk alternates between extended periods of stability and big jumps. They jitter too much during the stable periods and lag too much after the jumps. While dynamic mixes of EMAs perform much better, as we saw in Figure 4.6, they introduce turbulence, where T_{eff} plummets. Learning that excludes turbulence can never be fully efficient in dealing with big surprises.

Diffusions

In our previous models, θ usually shifts by at least 0.25 when it resets. This chapter explores the learning process when all θ changes are gradual. The classic example of gradual change is a random walk. I will model each step $\Delta\theta$ as a bell-shaped wobble with standard deviation $\delta \equiv \text{std}(\Delta\theta)$, except that it bounces off imaginary walls at 0 and 1. The limit as $\Delta\theta$ approaches zero is a diffusion between reflecting barriers.

The closest analogue among reset processes is reset from a uniform distribution with $\Lambda = 6\delta^2$, since it generates the same density of θ and the same average variance of $\Delta\theta$. Still, the two processes don't look nearly the same. Figure 8.1 displays 10,000 random draws on a reflected diffusion with $\delta = 0.01$ and another 10,000 draws on its reset analogue.

Ordinary observers don't see the contrasts in Figure 8.1 and can't readily distinguish continuous changes from a mix of no change and jumps. Out of respect for ignorance, let us initially assume that learners wrongly treat reflected diffusion as a reset process and apply TD3L, the three-layered Tiny Doubters that worked so well in Figure 7.5. To allow finer gradations, let us double the number of TDs on the first layer to 24 while keeping 12 TDs on the second layer. Let doubts range geometrically from 10^{-1} to 10^{-12} on each of those layers and set a doubt of 10^{-6} on the top layer. Figure 8.2 displays the tracking for 10,000 plays, after 20,000 precursor plays dilute the impact of our priors on the TDs.

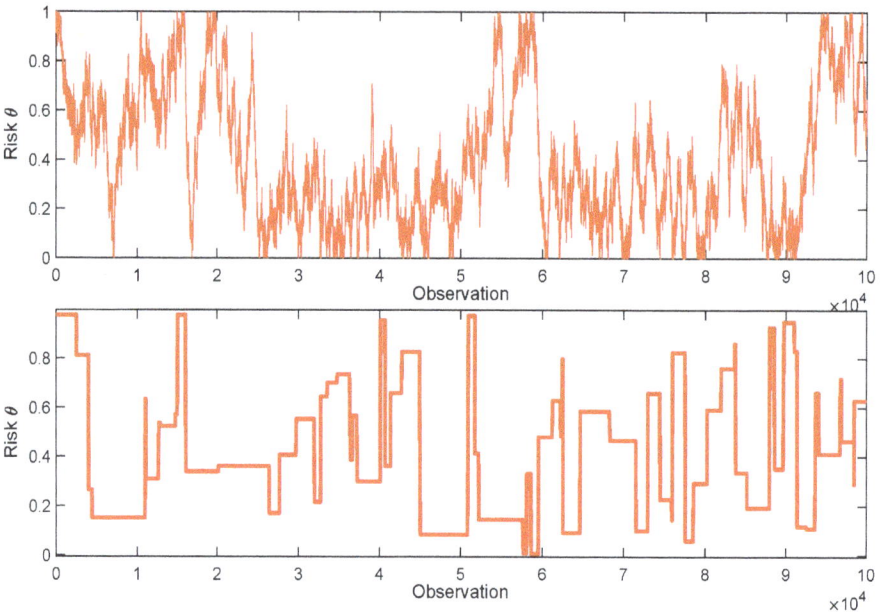

Figure 8.1: Sample risks for reflected diffusion with $\delta = 0.01$ versus reset from a uniform distribution at rate $\Lambda = 6\delta^2$

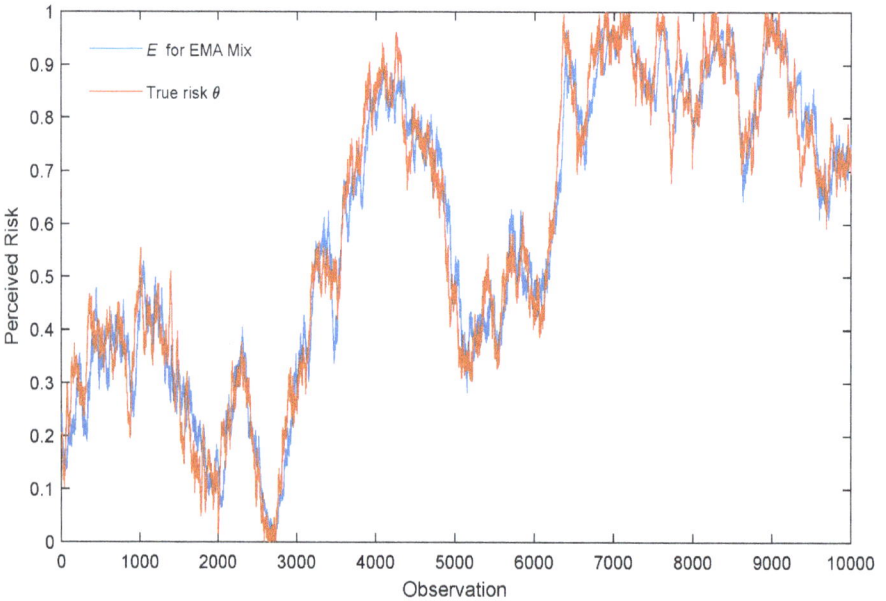

Figure 8.2: Mean beliefs and true risks for 10,000 plays of TD3L given reflected diffusion with $\delta = 0.01$

The tracking looks good overall but there are numerous false starts. On average, sharp spikes of 0.1 or more occur once every few hundred observations. Most spikes peak within 2-5 plays and vanish after 5-10 additional plays. They look like bursts of irrational anxiety.

Figure 8.3 charts the effective sample sizes for Figure 8.2. While most T_{eff} cluster between 50 and 200, with a median of 83, occasionally they dip much lower. Over a million plays, 8% of T_{eff} are less than 20 and 0.5% of T_{eff} are less than 5. The overall $\text{std}(\Delta E)$ is 50% higher than δ. That is, the mean belief E wobbles 50% more than the true risk θ. The entire excess coming from ΔE shifts of 5δ or higher.

The residual turbulence largely reflects the reliance on TDs, which are primed to look for large jumps. EMAs are better geared to analyzing a stable pace of change, thanks to their fixed learning rates. To test their impact, let's replace the 24 TDs on the first layer of TD3L with 50 EMAs, whose learning rates range geometrically from 0.0001 to 0.8. The TDs on the top two layers of TD3L stay the same. I call this EMA Mix. Its tracking is displayed in Figure 8.4 for the same plays and precursors as Figure 8.2.

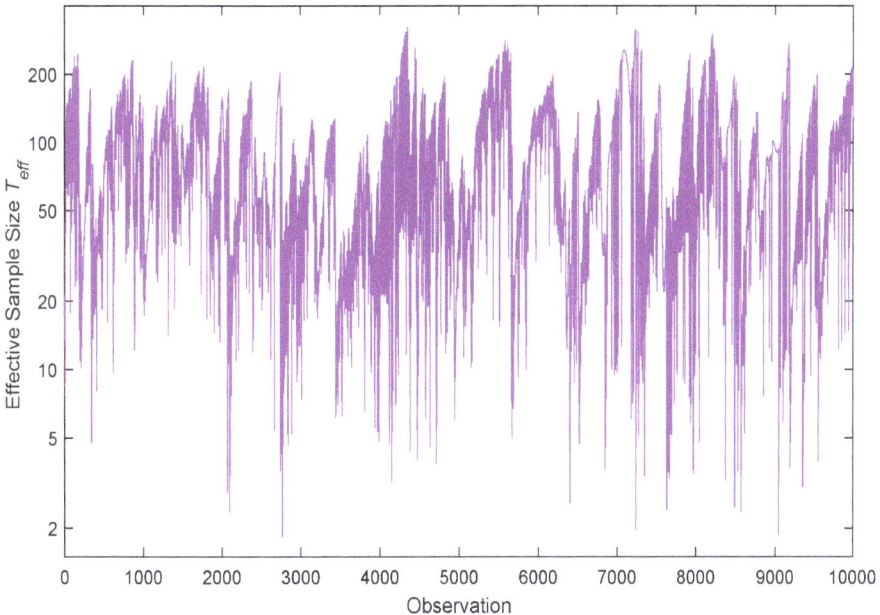

Figure 8.3: Effective sample size for Figure 8.2

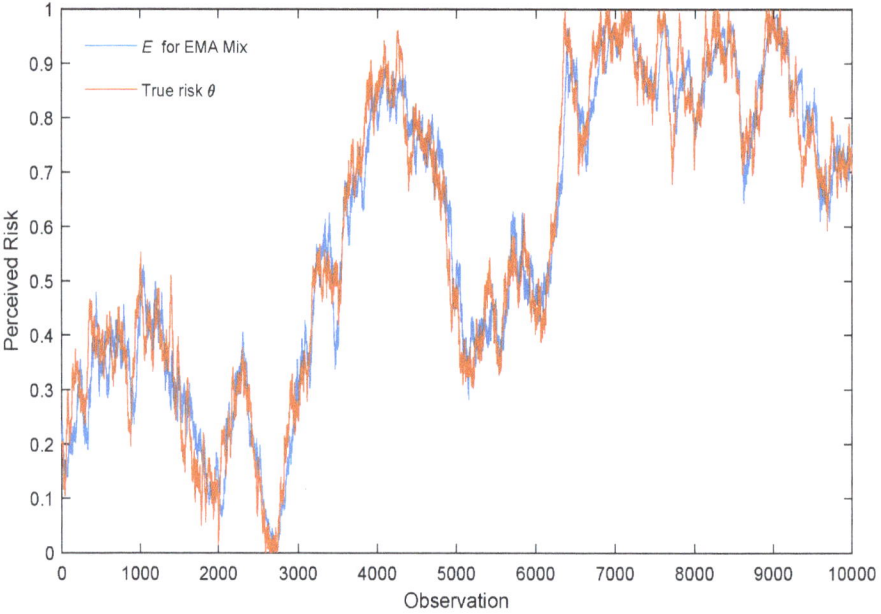

Figure 8.4: Mean beliefs for EMA Mix given the observations for Figure 8.2

EMA Mix offers similar core tracking to TD3L without the turbulent false starts and hence seems distinctly better overall. In a million plays, the RMSE is 0.063, nearly 0.01 less than for TD3L, and $\text{std}(\Delta E)$ drops to 8% below δ. As illustrated in Figure 8.5, learning rates are also strikingly more uniform. On average only 0.5% of T_{eff} fall below 20, while 90% cluster between 45 and 55.

The best EMA Mix boils down to a single EMA with duration near 50 except where E gets extremely low or high and the duration roughly halves. Here's why. For an EMA, $\text{std}(\Delta E)$ is close to σ_E / T_{eff}, which in turn should be close to δ for best tracking of the diffusion. Hence the best T_{eff} should be close to σ_E / δ. For most E, $\sigma_E \approx 0.5$, which implies $T_{eff} \approx 50$. However, σ_E plunges near extremes in E, in which case T_{eff} plunges too.

In principle, such analysis can help us find a good EMA Mix. In practice, it's not that important, as a good EMA Mix evolves quickly from a broad range of priors. If learners aren't sure an EMA Mix is appropriate, they can insert both TDs and EMAs at the lowest level and allow the higher TD levels to dynamically allocate between them.

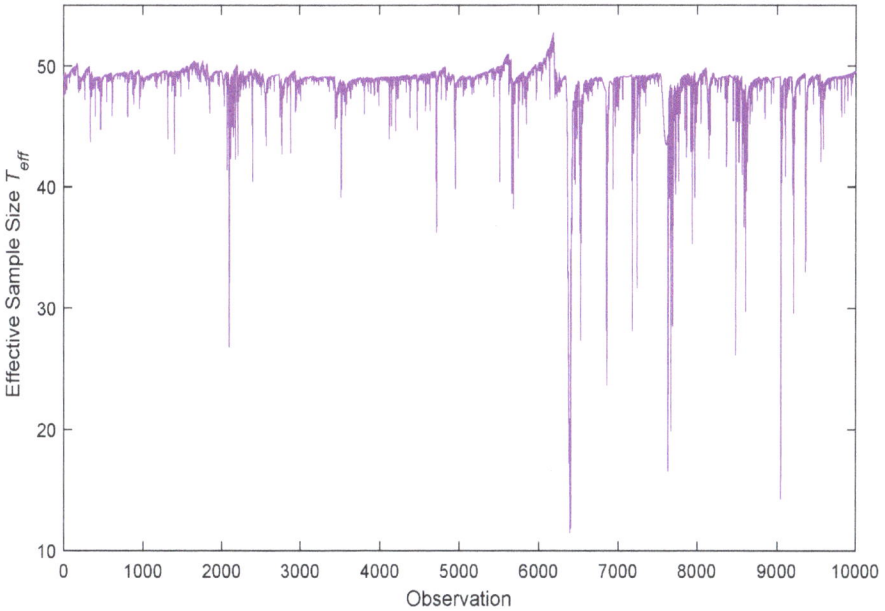

Figure 8.5: Effective sample size for Figure 8.4

For example, suppose we merge TD3L and EMA Mix, with 24 TDs and 50 EMAs at the base, 12 TDs above and one TD on top. Let's call this Hybrid Mix. It tracks diffusions nearly as well as EMA Mix does, by peeling most conviction away from the baseline TDs. For the parameters described earlier, EMAs typically dominate Hybrid Mix within a few dozen plays.

The greatest merit of Hybrid Mix is its versatility. Hybrid Mix tracks reset processes nearly as well as the best TD3L does, this time by peeling most conviction away from the baseline EMAs. For resets from a uniform distribution, it usually let TDs dominate after the first major jump. This makes Hybrid Mix a nearly perfect learner for either resets or diffusions, even though it doesn't know which process applies or the guiding parameters.

This illustrates a broader principle. For short-term prediction, flexible adjustment is more important than accurate model specification. However, specification can be crucial to long-term prediction, as we will see in Chapter 11.

Sources of Turbulence

Facing a pure reset process, Hybrid Mix alternates between calm and turbulent phases, as depicted in Figures 7.6 and 7.7. Facing a pure diffusion process, Hybrid Mix jitters regularly with few spikes, as depicted in Figure 8.5. Hence, neither instability in risk nor uncertainty about risk necessarily breed turbulence in rational learning. Why not? What features breed turbulence or inhibit it?

Analysis steers us to the cumulants of rational beliefs. From (3.3), the volatility of each cumulant is driven by the cumulant one order higher. In particular, the variance V, which from (6.4) is closely related to T_{eff}, bounces around with the skewness of beliefs. If T_{eff} hardly ever plunges to levels associated with turbulence, skewness must be consistently small. Invoking (3.3) again, we see that skewness can stay small only if kurtosis stays small, and so on up a never-ending cumulant ladder.

The only distribution with all higher cumulants zero is Gaussian (Marcinkiewicz 1938). It follows that rational beliefs must stay nearly Gaussian to prevent turbulence from emerging. In practice that is tantamount to keeping beliefs nearly symmetric. Asymmetry makes most odd cumulants non-zero, which in turn trigger shifts in even cumulants. The combined impacts resonate down to variance. To minimize asymmetry, never surprise beliefs with evidence of sudden, huge, systematic shifts in risk. Keep all change gradual and random.

Keeping change gradual and random is what a diffusion does. To the extent our beliefs remain Gaussian, rational learning couples a volatile ΔE with a relatively smoothly changing ΔV:

$$\Delta E \approx \frac{V}{\sigma_E^2}(x-E) \;\; ; \;\; \Delta V \approx -\left(\Delta E\right)^2 + \delta^2 \approx \frac{-V^2}{\sigma_E^2} + \delta^2 . \tag{8.1}$$

The negative component of ΔV reflects the extra clarity that new x provides. The positive component δ^2 matches the variance added by the next step of the diffusion. More generally, ΔV adds a third term in skewness times $\sigma_E(x-E)$. Absent skewness, V evolves toward $\delta\sigma_E$.

Multi-dimensional diffusions can be tracked in similar ways. Refined analogues to (8.1) are known as Kalman filters after their discoverer Rudolf Kalman (1960). Kalman filters have found many important applications. For example, suppose an anti-missile system is tracking an enemy missile. It takes noisy readings of the missile's latest location and uses them to update its estimate of the missile's velocity and future position. Here Kalman filters make excellent tracking tools because flight is a continuous trajectory and observations cluster near the mean. Since missiles heading east at 5 kilometers per second don't suddenly jump 50 kilometers west, any signal that suggests otherwise is flawed or points to a second missile.

Kalman filters update only means and variances because they presume skewness is always zero. They can go terribly awry if a boundary condition forces high skewness, as the following example shows. Suppose we are tracking θ for Bernoulli outcomes and apply (8.1) with $\sigma_E^2 = E(1-E)$ and approximations replaced with equalities. Eventually θ will stay near an edge long enough to squeeze σ_E^2 toward zero. Before long , this will drive V either above σ_E^2 or below zero, which in turn drives E outside the unit interval and onward to absurd extremes.

The simplest safeguard shifts the focus from V to the learning rate η in (6.4) and constrains its adjustment. Holding η constant creates an EMA. In general, Kalman filters function like flexible EMAs and work best when change is gradual. For more robust tracking, we can apply an array of Kalman filters with different parameters and allow the weights across the filters to adjust dynamically. Will tracking improve? Almost surely. Will tracking stay smooth? Not necessarily.

To see why good tracking can't always be smooth, imagine that a sophisticated missile creates diversionary trails. An array of Kalman filters will track that much better than a single Kalman filter. However, tracking improves less by avoiding big errors than by correcting them quickly. At one moment we're 95% confident that the missile is on path A rather than on alternative path B. Later our convictions reverse. The interim must include a period of confusion in which beliefs are highly sensitive to new data. If the two paths are sufficiently distinct, interim tracking is bound to look turbulent.

For more insight into tracking and turbulence, let us consider mixed processes in θ that reset with probability λ to a uniform distribution and otherwise diffuse with volatility δ. With $\lambda = 0.001$ and $\delta = 0.01$, EMA Mix and TD3L track θ nearly equally well for the same million plays, with RMSEs barely 0.2% apart. Figure 8.6 charts their respective predictions for 10,000 consecutive plays from that history. Clearly, EMA Mix generates some turbulence and TD3L generates more.

One measure of the ebbs and flows of turbulence is $\text{std}(\log T_{eff})$. Its value is 0.14 for EMA Mix tracking of the reflected diffusion, 0.44 for EMA Mix tracking of the mixed process and 0.82 for TD3L tracking of the mixed process. Figure 8.7, which charts T_{eff} for Figure 8.6, illustrates two different patterns of rational turbulence.

Figure 8.8 graphs the predictions of Hybrid Mix alongside the actual risk. While its spikes resemble those of TD3L, it is slightly less turbulent. Its RMSE is about 1.5% better than for either EMA Mix or TD3L and about 1.5% worse than for their simple average. A superior predictor might allow switches between various weighted averages of EMA Mix and TD3L.

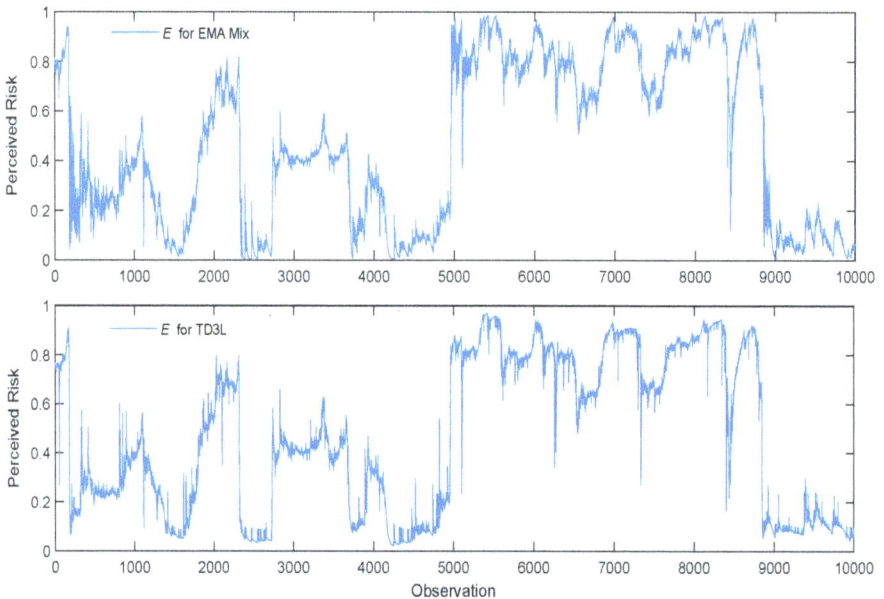

Figure 8.6: Mean beliefs for 10,000 consecutive plays of a mixed $(\lambda, \delta) = (0.001, 0.01)$ reset/diffusion as viewed by EMA Mix and TD3L

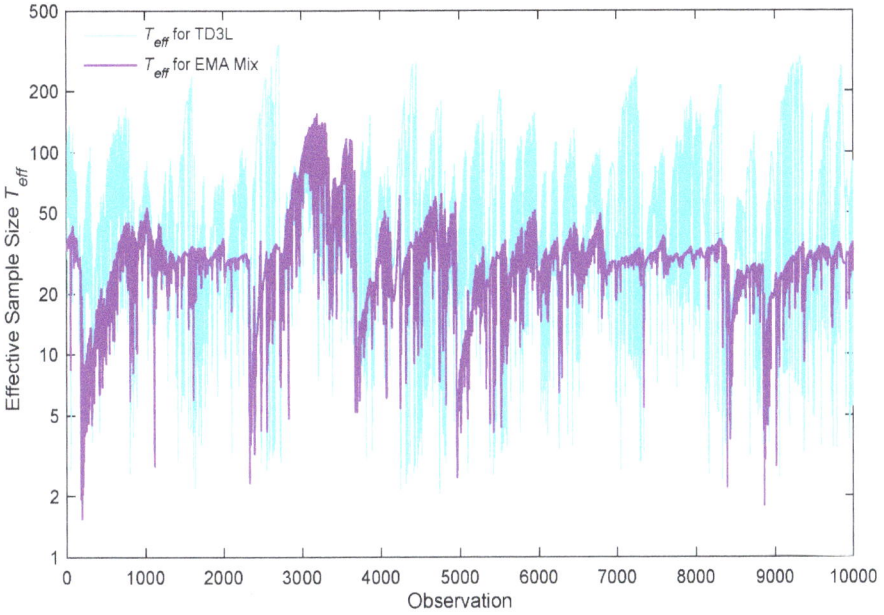

Figure 8.7: Effective sample size for Figure 8.6

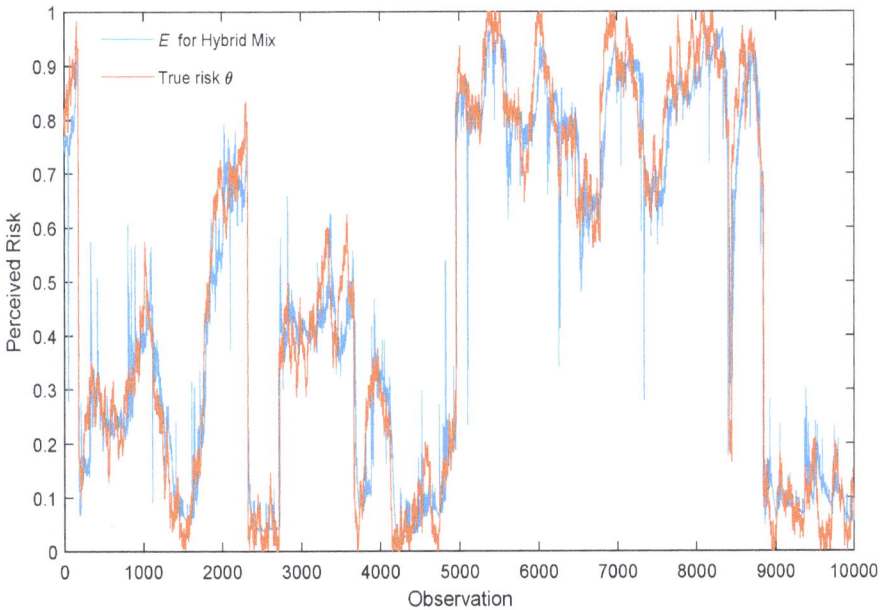

Figure 8.8: Mean beliefs for Hybrid Mix θ given the observations for Figure 8.6

To extend these findings to other mixes, note that multiplying λ and δ by the same factor leaves $\text{std}(\log T_{eff})$ roughly unchanged. Holding δ constant, higher λ accentuates turbulence while lower λ alleviates it. Suppose $\lambda = 10^{-4}$ and $\delta = 0.01$, which reduces λ/δ by 90% relative to our previous mixture. EMA Mix gets much calmer, its $\text{std}(\log T_{eff})$ halves to 0.21, and it outperforms TD3L in RMSE by 10%. This induces Hybrid Mix to emulate EMA Mix. Still, Hybrid Mix is chronically prone to turbulence with occasionally severe spikes, as Figure 8.9 shows.

Early discoveries of oxygen were misinterpreted due to the prevailing theory of phlogiston, an element purportedly shed by combustion, when in fact combustion is rapid oxidation. In finance, I think rational myopia has been misinterpreted as irrational exuberance. To demonstrate this more clearly, I need to fill a major gap. How confident can we be that markets of rational learners behave rationally? The next chapter will show that competitive pressures tend to make capital markets learn more rationally than their participants.

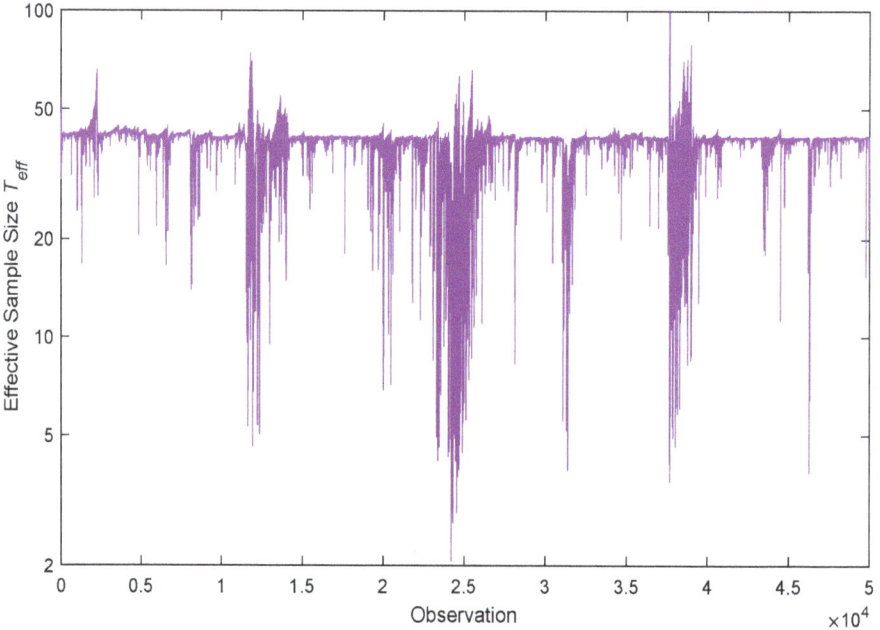

Figure 8.9: Effective sample size for Hybrid Mix in 50,000 plays with $\lambda = 10^{-4}$ and $\delta = 0.01$

9

Markets as Learning Machines

In standard finance models, the various agents i form identical probability assessments $p_i \equiv p$. These are known as homogeneous beliefs and imply parallel attitudes toward trading. Everyone wants to be long at prices higher than E and short at prices less than E. The only price that can clear markets with net holdings zero is E, which inclines agents to stick with what they have.

Differences in risk aversion and uneven changes in wealth justify only limited trading. Given homogeneous beliefs, every agent should hold risky assets in the same relative proportions. Most trading will simply shift the holdings of the composite risky asset as wealth or risk aversion changes. Neither can homogeneous beliefs explain the coexistence of major longs and major shorts. Even if many confused traders entered the market or seasoned traders lapsed into confusion, the market should restore its calm homogeneity once fools get parted from their money.

The only plausible explanation for why capital markets trade so much so often is that they are pervaded by disagreement. No analyst can sample future returns, know exactly what past information is relevant to current trends or anticipate all the ways current trends might change. Even ultra-calm, ultra-analytic traders will disagree about fair price and sometimes disagree a lot. However, orthodox finance models tend to exclude sustained disagreement. The investors with forecasts that most closely match the evidence win capital from the rest, making the market more homogeneous. Nobel Prize winners Lars Hansen (2007, 2010), Thomas Sargent (2008), and Joseph Stiglitz (2010) have each described this quandary as one of the central challenges of macroeconomics.

How can the capital market aggregate different individual beliefs? How can it let them compete vigorously without the best crushing the rest? How can heterogeneity naturally regenerate? Rational learning offers a straightforward, multi-part answer:

- When risks occasionally change without clear notice, uncertainty will ebb and flow, with tiny doubts occasionally fanning into big doubts and triggering rational turbulence.
- Tiny doubts will vary across different agents and occasionally fan into major disagreements, analogous to individual uncertainty.
- Between random noise and shifts in trends, the dominant beliefs and doubts will change over time, promoting heterogeneity.
- A market of rational gamblers tends to behave like a single rational mind, with individual views weighted by the capital backing them.

To place market reactions in context, let's return to Harold's Surprise. After 10,000 plays with $\theta = 0.5$, reds start occurring at every play. Various bots evaluate the evidence as it unfolds and adjust their forecasts of θ. How might they form a consensus?

Let's start with a standard voting mechanism and see how it fares. Suppose that just before each play, pairs of forecasts vie for bot votes until one forecast F beats all others. If every bot votes for the F closest to its own E, the median E is guaranteed to win. This is known as the Median Voter theorem.

We can generate the same result by announcing a tentative price F for a lottery ticket that pays the next outcome x, letting each bot choose to be long or short one ticket, and iterating until total longs match total shorts and the notional market clears. If we allocate TD doubts equally across $\{10^{-2}, ..., 10^{-16}\}$, the median doubt of 10^{-9} will always win. Figure 9.1 displays the median E trajectories for 200 different simulations of Harold's Surprise with $x_0 = 0$. Hardly any trigger much change in the median before the 25th straight red.

An alternative selection method equates the consensus F to the simple average E. This is equivalent to a market in which each bot bids in proportion to its expected profit $E - F$ on each lottery ticket and F adjusts to clear the market. Figure 9.1 displays those trajectories too. While the average responds sooner to new information than the median does, eventually it lags behind.

Would that we could form a weighted-average consensus and let the weights adjust dynamically to favor the better forecasts. It would be even better if we could do this without systematically misjudging the influence of blind luck. Remarkably, this is much easier to achieve than it sounds. All we need to do is:

i. Treat each bot as a gambler seeking to maximize long-term wealth by buying or selling lottery tickets out of its capital.

ii. Spread capital sufficiently widely that no single bot believes it significantly influences the ticket price F.[2]

Figure 9.1: Median and mean beliefs for 200 simulations of Harold's Surprise with $x_0 = 0$, given 15 Tiny Doubters with $\lambda = \{10^{-2},...,10^{-16}\}$

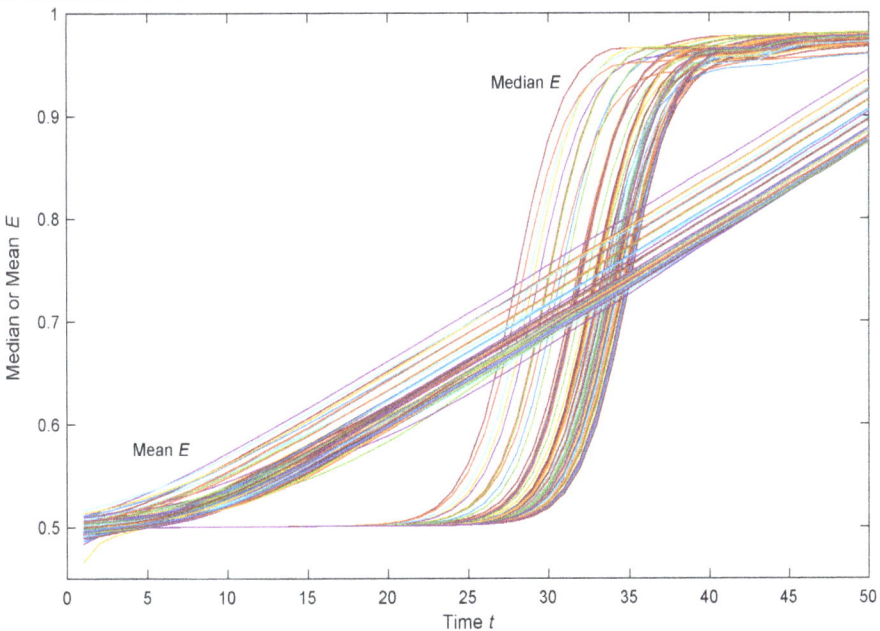

iii. Let F adjust until the bidding market clears and then treat it as the consensus forecast.

iv. After each outcome x is observed, adjust each bot's capital for the profit or loss and bet on the next x.

Since all this follows from the Invisible Mind theorem, let's review its derivation. To satisfy (i), gamblers should aim to maximize expected log returns and apply the Kelly criterion (4.1). Given (ii) and payoffs of 0 or 1, gambler i with mean belief E_i and share k_i of total capital K should buy $k_i K(E_i - F)/(1-F)$ tickets. For total bids to clear the market, as required by (iii), F must equal the capital-weighted average $\sum_i k_i E_i$. Applying (iv), the outcome changes capital shares by

$$\Delta k_i = k_i \frac{(E_i - F)(x - F)}{F(1 - F)} \tag{9.1}$$

which matches (4.2) for $\sigma_F^2 = F(1-F)$. I display the formula again in order to highlight a crucial correspondence. Suppose an investor believes that each bot forecast is correct with likelihood that matches its capital share. The Bayesian update is given by (9.1).

We can express this informally in various ways:

- Competitive Kelly-driven markets behave like rational analysts testing hypotheses against the data.
- An analyst applying Bayes' rule acts like an efficient market in rival beliefs.
- Capital shares represent the market's views on credibility.
- Subjective probabilities behave like credibility capital.

Any learning mechanism that satisfies properties (i)-(iv) can be viewed as a kind of market overlay, whether or not it explicitly operates a market. Figure 9.2 displays market prices for 200 different simulations of Harold's Surprise with $x_0 = 0$, given 16 TDs with $\lambda_i = 10^{-i}$ and initially equal capital.[3] The core transitions start much sooner than for the median and they end much sooner than for the simple average.

Figure 9.2: Market price for 200 simulations of Harold's Surprise with $x_0 = 0$, 16 TDs with $\lambda_i = 10^{-i}$ and equal initial capital

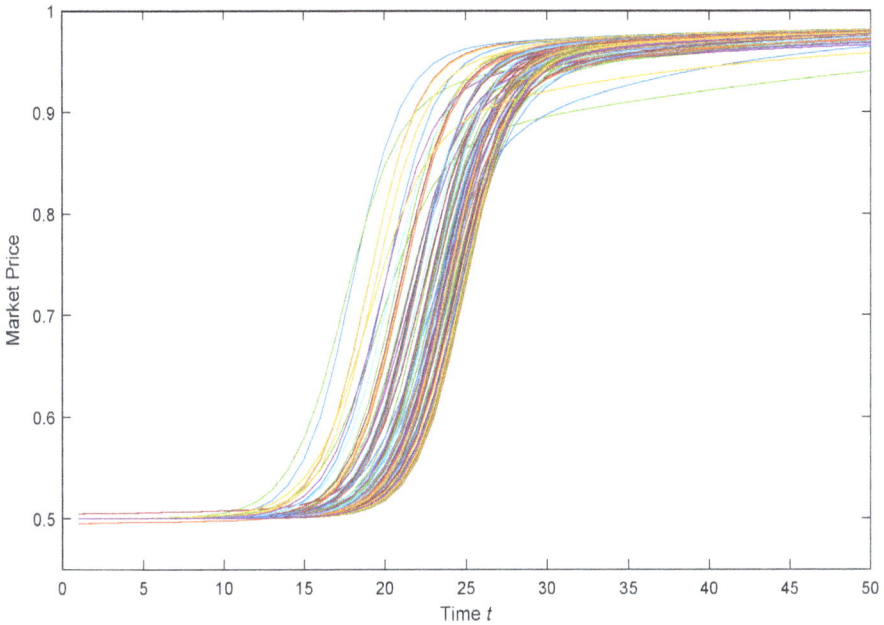

Shifts in capital explain the quicker response. Figure 9.3 charts the average capital shares. The left edge shows the shares at time 0, after 10,000 plays at $\theta = 0.5$. TDs 1-3 lost so much capital they aren't visible. Anticipating on the order of 10 to 1,000 shifts in θ, they bet too much on bogus changes. TD 4 anticipated a single shift on average, which wasn't wildly wrong, but it gambled too much and lost credibility. The remaining bots weighted the observations so similarly that TD 16 gained only 10% more capital than TD 5 and only 1% more capital than TD 6.

Market prices hardly budge for the first ten straight reds, as TDs 1-3 respond but never regain much influence. After a dozen straight reds, TD 4 gets noticeably bolder and starts to make an impact. On average, by $t = 16$ it restores its initial capital share of 6%; ten more reds boost its share to 75%. Meanwhile, TD 5 gets more confident about the switch and starts to share the gains. After thirty straight reds, TD 4 owns nearly 80% of the capital and TD 5 owns 90% of the rest.

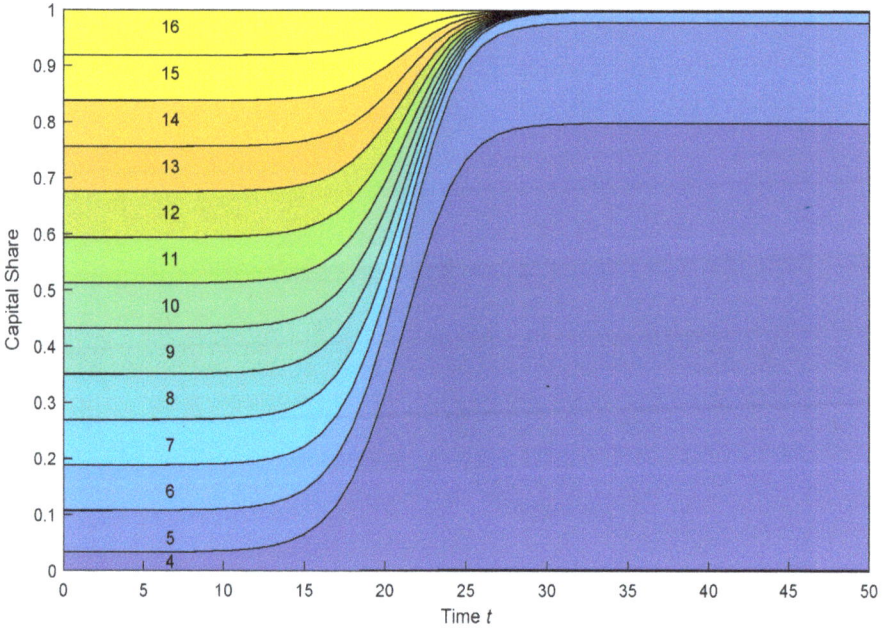

Figure 9.3: Average capital shares in Figure 9.2

Inter-Threaded Markets

The insignificance in Figure 9.3 of TDs 1-3 indicates a shortcoming of the market. If the straight reds were preceded by millions of fair plays, TDs 4 and 5 would lose nearly all their capital too and the market would need at least 10 reds more to respond to Harold's Surprise. In general, a market wedded to stable strategies can't respond quickly to their failure.

The easiest way to counteract this is to ensure each bot a minimum share in aggregate capital. Perhaps different bots exchange smidgeons of their views. Perhaps they have small equity stakes in other bots' wealth. Perhaps the state imposes a wealth tax-cum-transfer. The impact is most neatly modeled by positing a fifth property for our market:

v.   At the end of every round, nudge each capital share a fraction λ_{AGG} toward equality or some other long-term norm.

I will call the combination (i)-(v) an inter-threaded market or ITM. I say inter-threaded rather than interwoven to emphasize that λ_{AGG} is typically dwarfed short-term by the adjustments in (9.1). The threading need not be uniform across bots or constant over time although I assume that here for convenience. It just needs to keep bots from being squeezed to near-permanent irrelevance. The similarity between tiny redistribution across the aggregate market and tiny doubts across individual beliefs is why I use the same letter λ to describe both.

It is very hard to foresee which λ_{AGG} is best. A robust alternative imposes multiple ITM overlays, each for a different λ_{AGG}. This generates multiple market prices, on which we can impose yet another ITM overlay to form a higher-level consensus. Mathematically, these layers of ITMs behave like the layers of TDs introduced in Chapter 7. The only difference is their driving mechanism. TDs update the judgments of a careful analyst on the credibility of various hypotheses. ITMs are markets of gamblers who seek to profit at others' expense.

The isomorphism has two profound implications. First, the human mind might operate in far more market-like ways than we realize. Perhaps what we describe as self is a forum of competing internal interests, with decisions that resemble consensus prices.[4] Second, markets can rectify some human foibles. To take an extreme example, imagine a market of stubborn dogmatists. Each gambler i is appallingly certain that the risk is θ_i and never changes. Fortunately, a tiny share λ of wealth is redistributed every period. That suffices to turn the market into a Tiny Doubter, which adapts flexibly to new information.

For a more nuanced example, imagine each trader learns at a fixed rate and hence behaves like an EMA. As Figure 9.4 shows, no EMA responds to Harold's Surprise with the S-shaped behavior we know is optimal. Moreover, the short-duration EMAs that respond quickest to all-reds are in most contexts wildly noisy. Suppose 50 EMA traders with durations $T_i = 10^{i/10}$ spaced geometrically between 1 and 100,000 are embedded in an ordinary market with $\lambda_{AGG} = 0$. In 10,000 plays with $\theta = 0.5$, long-duration EMAs will so dominate that even 50 straight reds will rarely push the market price as high as 0.55.

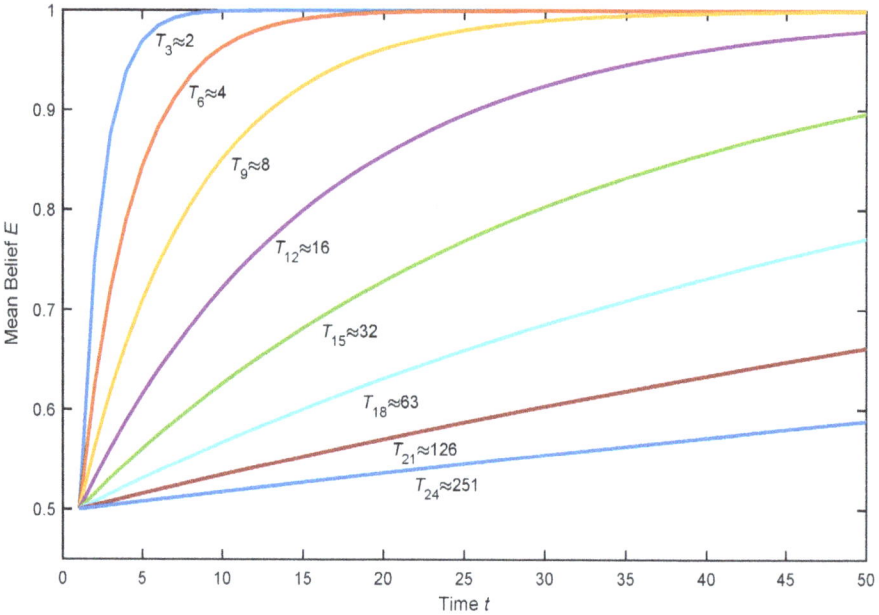

Figure 9.4: Mean beliefs for EMAs observing Harold's Surprise

Modest inter-threading can fix that. When $\lambda_{AGG} = 10^{-6}$, net transfers are tiny. After 10,000 plays with $\theta = 0.5$, the total capital share on $T_i \leq 10$ is only 0.0004%. Still, this is over 10^{100} times the corresponding share when $\lambda_{AGG} = 0$ and triggers transitions that resemble Figure 9.2. The market price typically surges by 0.4 in seven plays around $t = 20$ and reaches 0.99 by $t = 30$. The key driver, depicted in Figure 9.5, is a huge shift in weight toward short-duration EMAs.

To improve this further, we can organize ITMs of EMAs for various λ_j and embed them in a higher ITM. Figure 9.6 displays the core structure, which mimics EMA Mix in Chapter 8. Figure 9.7 displays the market price when the top two layers set $\lambda_j = \{10^{-1}, 10^{-2}, \ldots, 10^{-8}\}$ and $\lambda_{AGG} = 10^{-8}$. The core transitions start sooner than for the TDs in Figure 9.2, finish sooner, and converge faster toward $\theta = 1$. They don't resemble any of the EMA paths in Figure 9.4.

Figure 9.5: Average capital shares for Harold's Surprise with $x_0 = 0$, 50 EMAs with $T_i = 10^{i/10}$, and $\lambda_{AGG} = 10^{-6}$ across EMAs

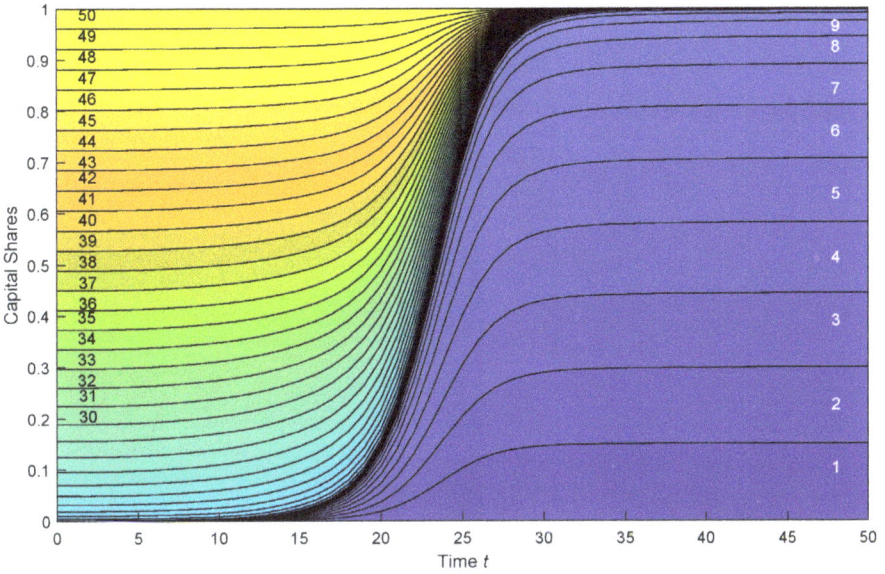

Figure 9.6: Learning with two layers of ITMs on top of EMAs

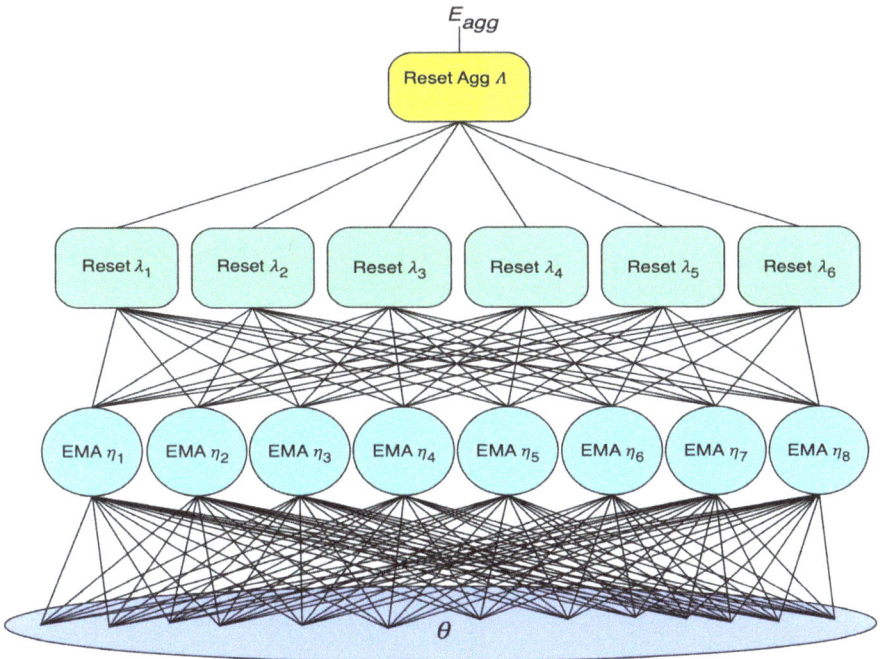

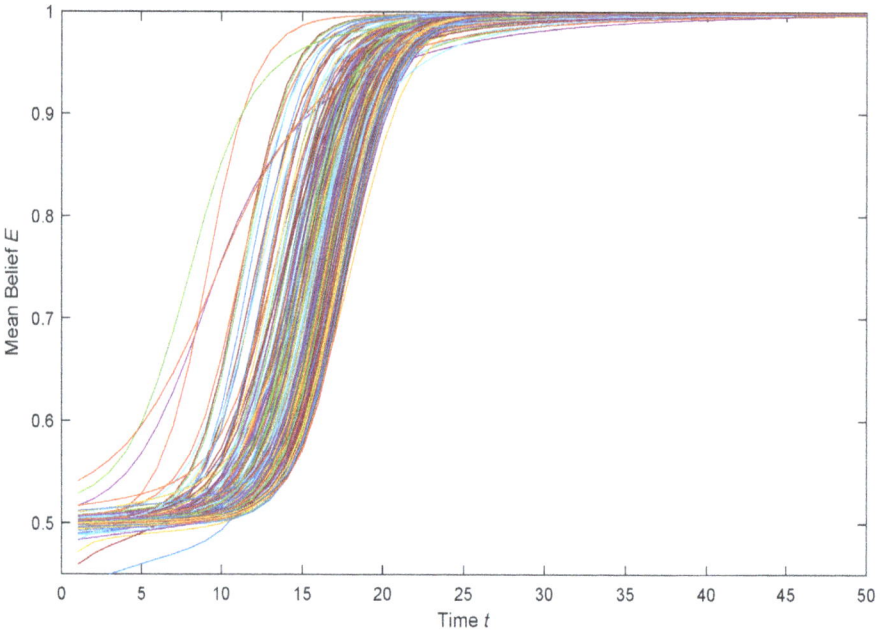

Figure 9.7: Price for Harold's Surprise with $x_0 = 0$, 50 EMAs with $T_i = 10^{i/10}$, 8 ITMs across EMAs with $\lambda_j = 10^{-j}$, and $\lambda_{AGG} = 10^{-8}$ across ITMs

Philip Tetlock and Dan Gardner (2015) coined the word "superforecasters" to describe people who forecast the way they optimally should. Superforecasters apply Bayes' rule to relevant data and allow for a wide range of uncertainty. Tetlock's empirical research suggests that most self-styled experts get too fixated on a favored trend or claims that garner attention to forecast well. In that sense they resemble fixed-θ predictors or pure EMAs. Figure 9.7 shows that capital markets can behave like super-forecasters even when its participants don't.

Let me close this chapter by tying up some mathematical loose ends. My derivation of the Invisible Mind theorem was limited to single-layer learning. Below I extend this to multiple layers and also justify the Kelly criterion for maximizing log wealth. Let p_i denote gambler i's beliefs. The market's aggregate beliefs are $p_{AGG}(\theta) = \sum_i k_i p_i(\theta)$ with mean $E_{AGG} = \sum_i k_i E_i$. Bayesian updating revises them to

$$p_{AGG}^*(\theta) = \begin{cases} \dfrac{\theta}{E_{AGG}} p_{AGG}(\theta) & \text{if } x = 1 \\[2ex] \dfrac{1-\theta}{1-E_{AGG}} p_{AGG}(\theta) & \text{if } x = 0 \end{cases}. \tag{9.2}$$

To show that a decentralized, layered approach generates the same result, let us compute the number W_i of securities that gambler i should hold assuming a price of F. Expected log growth

$$E_i \log(k_i K - W_i F + W_i) + (1 - E_i) \log(k_i K - W_i F)$$

is maximized when $W_i = k_i K (E_i - F) / F(1 - F)$, as claimed in (4.1). To clear the market, $\sum_i k_i W_i = 0$ which implies $F = E_{AGG}$.

Next let's examine the payoffs. If $x = 1$, gambler i gains $(1 - E_{AGG}) W_i$ for a new capital share $k_i^* = k_i E_i / E_{AGG}$ and new beliefs $p_i^*(\theta) = p_i(\theta) \theta / E_i$. If $x = 0$, gambler i loses $E_{AGG} W_i$ for a new capital share $k_i (1 - E_i) / (1 - E_{AGG})$ and new beliefs $p_i(\theta)(1 - \theta) / (1 - E_i)$. This implies

$$k_i^* p_i^*(\theta) = \begin{cases} \dfrac{\theta}{E_{AGG}} k_i p_i(\theta) & \text{if } x = 1 \\[2ex] \dfrac{1-\theta}{1-E_{AGG}} k_i p_i(\theta) & \text{if } x = 0 \end{cases}.$$

Summing this over all gamblers matches (9.2). Extension to three or more layers is straightforward.

10

Social Pressures

Let's face it: casino gambling is weird. It lures players who aim to profit from their bets, who know that gamblers on average lose money, and who cannot on average do better than average. Governments are torn between trying to protect gamblers from their profligacy and trying to profit from it through taxes and state-run lotteries.

Even for purely rational gamblers, there is an inherent tension between individual beliefs and market outcomes. Given a mean belief E_i and a market price F that equilibrates at the aggregate E, a Kelly bettor i with one unit of capital expects a mean profit of $(E_i - E)^2 / E(1 - E)$. Aggregating over all bots, the aggregate expected profit per unit of capital is $V/E(1-E)$. That's exciting ex ante. Since the perceived increment equals the Bayesian learning rate η absent doubts and anticipation, we can estimate its value from price movements. Yet when the bots bet solely against each other, the aggregate net return must be zero. That's disappointing ex post.

We can summarize this as

$$\text{Excitement} = \frac{V}{E(1-E)} = \text{Disappointment} \approx \eta. \tag{10.1}$$

For the market to work efficiently, each gambler bot needs to reject the irony implicit in (10.1). It must believe that others are over-optimistic on average by more than the disappointment, so that they stand to lose in aggregate exactly what the gambler expects to win.

This requirement is less restrictive than it appears. Suppose you and I are the only bettors in the market, with equal capital. My private analysis suggests $E_{me} = 0.3$ while yours, judging from your initial bid, suggests $E_{you} = 0.7$. Since I respect your analytic skills as much as mine, I revise my bid to $E_{me}^* = \frac{1}{2}(E_{me} + E_{you}) = 0.5$ and the market clears at $E = 0.6$. Does this make our market inefficient? Not at all. We just need to treat E_{me}^* as my relevant mean belief. If you reweight your bid to $E_{you}^* = \frac{3}{4}E_{you} + \frac{1}{4}E_{me} = 0.6$, that's ok too and the market clears at $E = 0.55$ In a huge market, where we know few of the participants, the core social pressures concern the aggregate signal E. The rest of this chapter will assume that this is the only information that is broadly shared.

The first social pressure I will address is groupthink, modeled as a tax on doubt. At every period for every bot, a fraction τ of every conviction $p(\theta)$ gets transformed into conviction for the current E. Groupthink convictions on previous E get taxed as well. Low τ turns out to have little impact, as old groupthink clashes with new groupthink and helps sustain learning. In contrast, high τ eventually crushes variance and freezes the market.

To illustrate this, let's posit reset at rate $\Lambda = 0.001$ from a uniform distribution and a baseline market that mimics a Perfect Learner. For each simulation, l conduct 200,000 plays with a fixed τ and measure the RMSE over the last half. Figure 10.1 charts the scatterplot of RMSEs versus τ for a thousand simulations: 20 each for 50 τ spaced geometrically from 0.0002 to 0.5.

In this model, groupthink has negligible impact for τ below the reset rate λ and modest impact even for τ twice as high. Tracking markedly deteriorates thereafter. By $\tau = 0.1$ the market E is bound to freeze, with RMSE ranging from 0.29 for freezes near the center to over 0.5 for freezes near a side. Figure 10.2 displays a sample path for 20,000 observations on risk, baseline price and groupthink-constrained price under an intermediate tax of 0.02. Groupthink restrains turbulence at the expense of tracking.

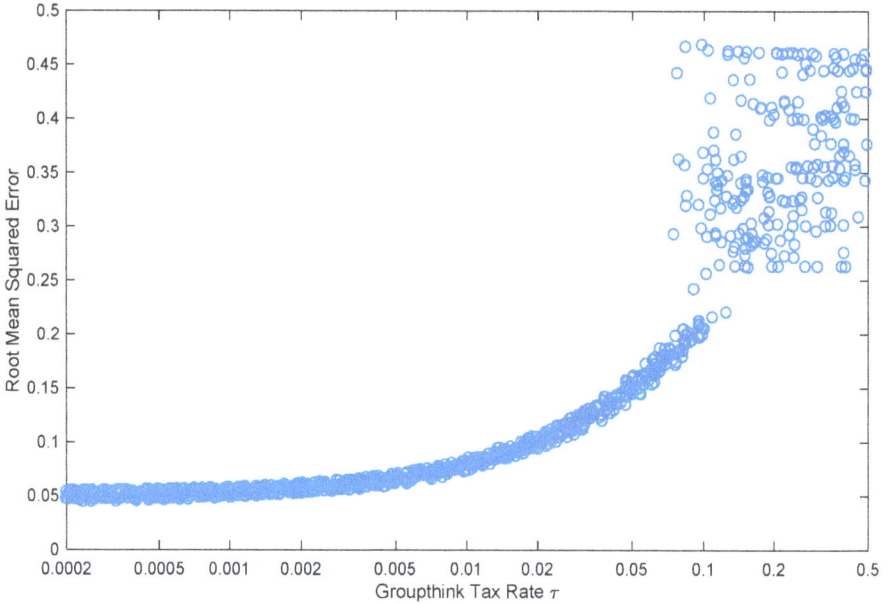

Figure 10.1: Scatter plot of 1,000 RMSEs for various groupthink tax rates τ

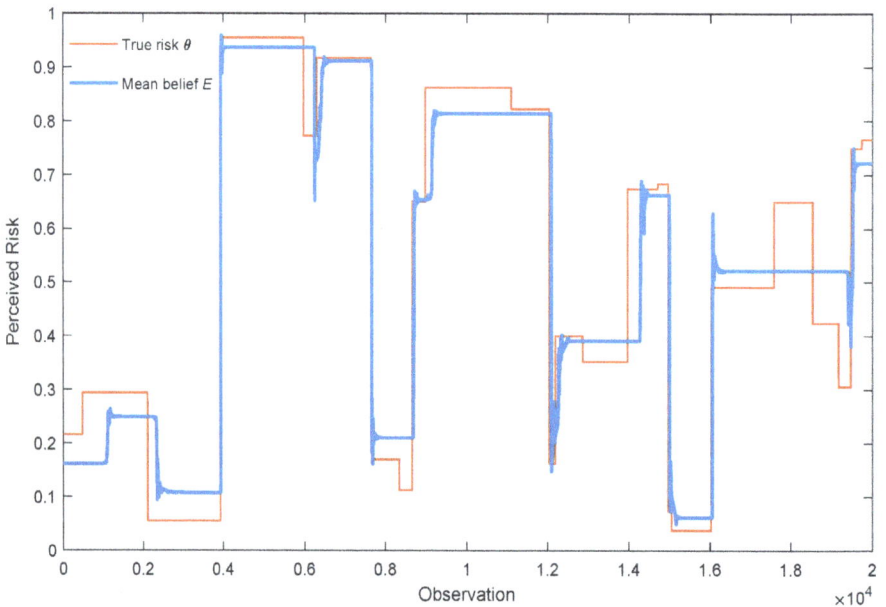

Figure 10.2: Sample path for 20,000 observations on risk, baseline price and groupthink-constrained price with $\tau = 0.02$

Price adjustment noticeably lags risk and usually freezes short of the goal, until a "risk-quake" triggers a brief, mildly turbulent adjustment.

Let us now consider a model of caution. While no bot changes its own views, each acknowledges that it has been too optimistic about its betting prowess and shrinks its bet to a fraction f of full Kelly. We can measure its deference or herding as the relative shrinkage $1-f$. From our modeling perspective, the motives are irrelevant.

In fact, shrinkage can occur without any social adjustments. Suppose the bot with wealth K_i aims to maximize $\left\langle -K_i^{1-\gamma} \right\rangle$ for some $\gamma > 1$. This utility weighting, which implies constant relative risk aversion γ, is effectively $\left\langle \log K_i \right\rangle$ for γ near 1 but penalizes large losses relatively more as γ increases. To first approximation, the optimal Kelly bet shrinks by a factor $1/\gamma$, which likens relative risk aversion γ to deference $1-1/\gamma$.

As noted in Chapter 4, a consistent application of this approach is known as a fractional Kelly strategy. Professional gamblers highly respect it, at least down to half Kelly or quarter Kelly, because they can expect to earn nearly as much money in expectation as full Kelly with much less risk of huge losses. In the limit of tiny bets on tiny payoffs, full Kelly gamblers face a risk $1-L$ of a relative peak-to-trough drawdown greater than L. Fractional Kelly gamblers reduce the drawdown risks to $(1-L)^{2/f-1}$ (Chin and Ingenoso 2007).

Figure 10.3 shows that the reductions are substantial. While enrichment slows, it slows by considerably less than the $1-f$ shrinkage in bets. Specifically, the mean logarithmic rate of growth declines by $(1-f)^2$, which amounts to 25% for half Kelly and 56% for quarter Kelly.

To test the market impact of fractional Kelly, I compared prices for two million plays assuming that θ resets at rate $\lambda = 0.001$ from a uniform distribution. Here are the main findings:

- **Half Kelly looks a lot like full Kelly.** For an optimal TD learner or market of TDs, half Kelly RMSE exceeds full Kelly RMSE by 6%. For markets of EMAs, the gap is only 3%. For TDs, 87% of half Kelly E forecasts lie within 0.02 of full Kelly E.

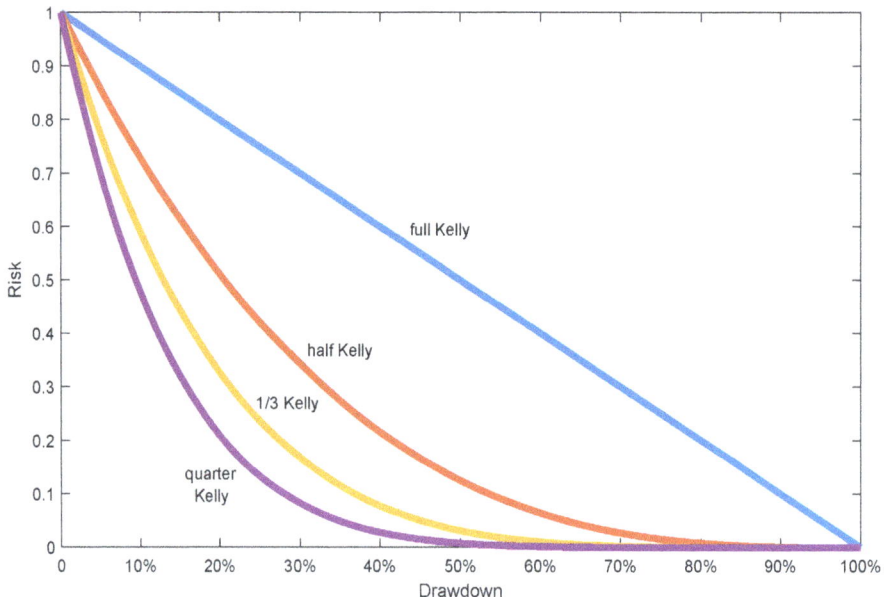

Figure 10.3: Drawdown risks under fractional Kelly

- **Quarter Kelly on TDs looks a lot like full Kelly on EMAs.** Their RMSEs nearly match, as do the distributions of mean absolute errors $|E - \theta|$ and mean absolute price changes $|\Delta E|$.
- **Low Kelly coarsens average tracking. For markets of TDs,** $|E - \theta|$ averages 39% higher for quarter Kelly and 15% higher for half Kelly than for full Kelly.
- **Low Kelly adds jitter.** For TDs, $|\Delta E|$ averages 44% higher for quarter Kelly and 25% higher for half Kelly than for full Kelly.
- **Low Kelly prolongs turbulence.** Learning rates η for TDs exceed 0.03 for 9% of plays under full Kelly, 13% under half Kelly and 15% for full Kelly.
- **Low Kelly dampens extreme turbulence.** Learning rates exceed 0.2 for 0.5% of plays under full Kelly, 0.2% under half Kelly and never for quarter Kelly.

Figure 10.4 charts the relative increase in RMSE as the Kelly fraction declines. The increase is roughly cubic in $1 - f$. It is minor above half Kelly but rises quickly below quarter Kelly.

Figure 10.4: Increases in Kelly-based RMSEs relative to optimal

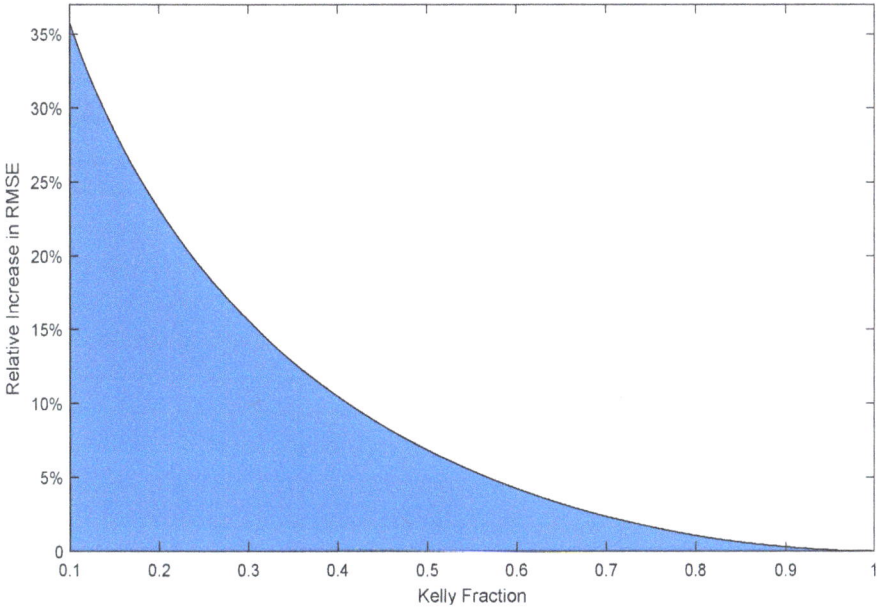

Figure 10.5 compares a sample 10,000-play price path for full Kelly versus quarter Kelly. The underlying servicing observations are the same for both price paths as are the servicing observations for 50,000 plays prior. The only differences are the Kelly reactions. Full Kelly tracks well overall at the cost of occasionally intense false starts, which look like solitary spikes in Figure 10.5 although they typically last a dozen plays or more. Quarter Kelly eliminates the longest spikes but is choppier than full Kelly and adjusts more slowly to changes in risk. To observers unaware of rational turbulence, quarter Kelly will appear more consistently rational, whereas full Kelly peppers calm with occasional panic.

Deference and risk aversion seem so endemic that perhaps markets aren't rationally turbulent enough. Reports of flash turbulence on stock markets triggered by program trading may signal the progress of AI as much as its shortcomings. Chapter 4 noted that aggressive betting helped an AI poker player defeat the world's top pros. Evidently, we humans don't just lack the brains to beat AI gamblers long-term. Our stomachs won't handle it either.

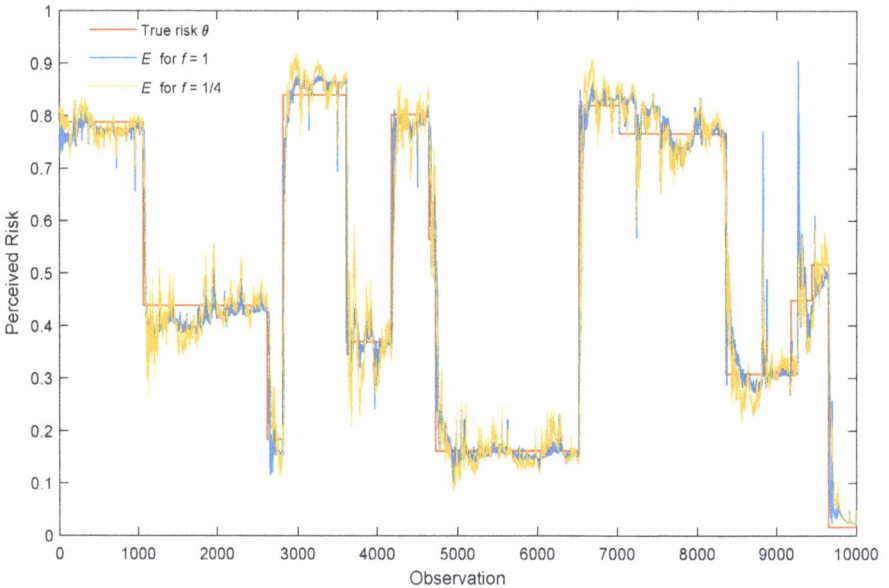

Figure 10.5: Sample 10,000-play price path for full Kelly and quarter Kelly

Collaborative Learning

Our treatment has assumed that every market learner processes the same core information, forms completely independent judgments and tempers them only at the end. In reality, no single brain can gather all the available information, let alone process it well. To expand our range, we draw on information from others, which mingles facts, fictions, insights and prejudice. Raw observations often resemble the proverbial reports from blind men touching parts of an elephant. Higher layers of analysts distill, reinterpret and synthesize those reports. The market weights their judgments by perceived credibility, aka reputation.

Weighting opinion by reputation can encourage mental laziness. Imagine people who always believe the market E is the true mean. They investigate nothing, they wager nothing, they lose nothing: fractional Kelly with a fraction zero. If everyone did this, no one would ever learn except through rude awakenings. Fortunately, capital markets offer incentives to distinguish useful reputations from useless ones. Many reputable investors have little direct knowledge of what they invest in but have identified good experts they rely on. Those experts in turn tap more specialized expertise of others.

If each player gets to access private data z_i, how can we best use that information to improve our estimation? Suppose θ is a known function ϕ of $y = \mathbf{z}\boldsymbol{\beta}$, where $\mathbf{z}$ denotes a row vector of the z_i and $\boldsymbol{\beta}$ denotes a column vector of unknown coefficients β_i. Our Bayesian updating rule extends naturally to

$$\Delta p(\boldsymbol{\beta}) = p(\boldsymbol{\beta}) \frac{\phi(\mathbf{z}\boldsymbol{\beta}) - E}{E(1-E)} (x - E) \quad \text{for} \quad E = \langle \phi(\mathbf{z}\boldsymbol{\beta}) \rangle .$$

Unfortunately, this brings the curse of dimensionality. Allowing for 10 different values of each β_i over 10 different i generates 10 billion distinct values of $\boldsymbol{\beta}$. To make updates tractable, we need to focus on summary measures like the mean beliefs $\boldsymbol{\mu} = \langle \boldsymbol{\beta} \rangle$ and the value $\phi(\mathbf{z}\boldsymbol{\mu})$ at the mean. The simplest fix presumes a logistic relation

$$\theta = f_L(y) = \frac{1}{1 + e^{-y}} ,$$

which makes y the log odds of θ. Figure 10.6 depicts the logistic along with two clipped linear approximations. The best linear approximation in the central region has slope 0.25. A central slope of 0.178 minimizes the maximum deviation.

The main reason for preferring the logistic to clipped linear functions is that it simplifies the updates of $\boldsymbol{\mu}$ to approximately

$$\Delta \mu_i \approx \sum_j \text{cov}(\beta_i, \beta_j) z_j (x - \phi_L(\mathbf{z}\boldsymbol{\mu})) ,$$

where cov() denotes covariance. This follows from replacement of ϕ_L with a Taylor expansion in its partial derivatives. If the z_i are uncorrelated, we can further simplify this to

$$\Delta \mu_i \approx \text{var}(\beta_i) z_i (x - \phi_L(\mathbf{z}\boldsymbol{\mu})) . \tag{10.2}$$

Figure 10.6: Logistic curve and two clipped linear approximations

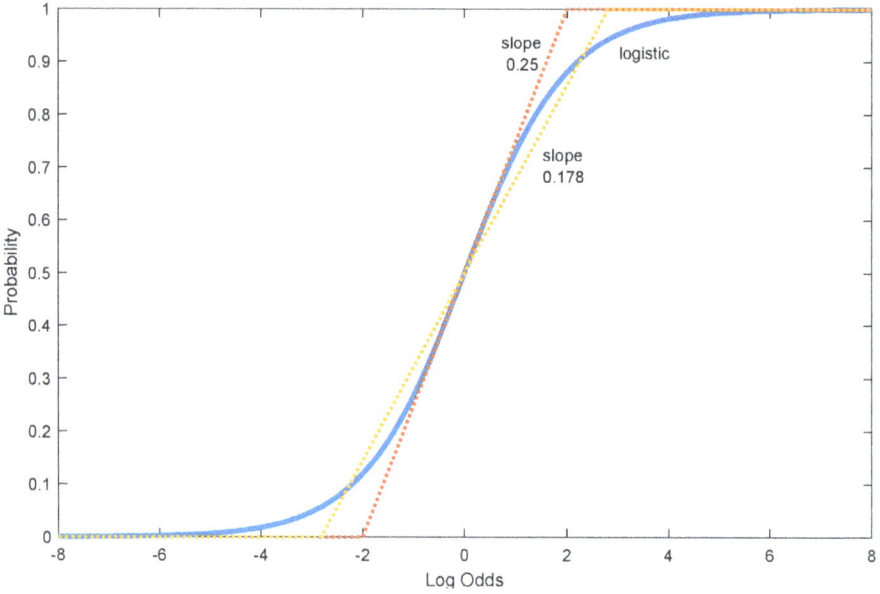

The general approach is known as Bayesian logistic regression. For a market interpretation, suppose that bot ij has very narrow expertise. It accepts the consensus view μ on every component other than i, where it is convinced the true mean is $\mu_i + \delta_j$, and expects its own information to shift fair price by approximately

$$\delta_j \frac{\partial \phi}{\partial \beta_i} \approx \delta_j z_i \phi_L (1 - \phi_L).$$

Applying full Kelly, the bot wagers approximately a $\delta_j z_i$ fraction of its capital. If every gambler acts likewise, the shift in μ_i after x is paid out is approximately (10.2). Hence, the capital market can induce collaborative learning relatively efficiently.

While it is easier to implement logistic regression through a bot than through a market, the bot might not give a better answer. Performance depends on the proxies used for $\text{var}(\beta_i)$ and their updates. Typically the various z_i are rescaled to the same range or standard deviation and the adjusted $\text{var}(\beta_i)$ presumed equal, so that

$$\Delta\mu_i = bz_i\left(x - \phi_L(\mathbf{z\mu})\right). \tag{10.3}$$

Iteration starts with a high b and gradually reduces it over time t. If the convergence is poor, b gets bumped higher. Decisions about b are often ad hoc or linked to hyperparameters not tested against the data.

A market-like framework can make decisions about hyperparameters less subjective. For example, imagine that θ is resampled every period from a uniform distribution, which makes past observations of x irrelevant. Fortunately, information $\mathbf{z}$ is available on the powers of $2\theta - 1$. Specifically, $z_i = (2\theta - 1)^i$ for $i = 0,\ldots,7$. No one divines the exact relation $\theta = \frac{1}{2}(z_1 + 1)$. Instead, 25 bots jk fit (10.3) over the course of 100,000 plays, $b_{jk}(t) = 10^{-j}(t+5)^{-k/4}$ for j and k ranging from 0 to 4. Figure 10.7 displays a scatter plot of $(\theta, \phi_L(\mathbf{z\mu}))$ pairs for the last 5,000 plays and compares it to the ideal. The excellent fit shows that good prediction need not require good understanding.

However, ITMs of (10.3)-type bots don't always perform well. Figure 10.8 displays the analogue to Figure 10.7 when $\mathbf{z}$ consists of powers of θ or $\theta - 0.5$. High correlation worsens the fit for powers of θ while uneven scaling worsens the fit for powers of $\theta - 0.5$. Let me caution that the fits for any of these $\mathbf{z}$ vary significantly across simulations although I have tried to pick representative results.

Collaborative learning needs more sophistication to decorrelate data and to modify functional forms. The most important adaptation adds layers of logistic regressions, with the outputs of one layer providing inputs to the next. The ensembles are known as neural networks, thanks to their stylized resemblance to the biological neural networks that characterize animal brains.

While neural networks have many market-like features, analysis falls outside the scope of this book. A fuller treatment should apply a multivariate Kelly criterion (Osband 2005) and explore its analogies both to the classic Markowitz (1952) formula for portfolio optimization and to the multivariate version of the Core Learning equation. Multivariate criteria would also help tighten the treatment in the next chapter on the relations between prices of present and future claims.

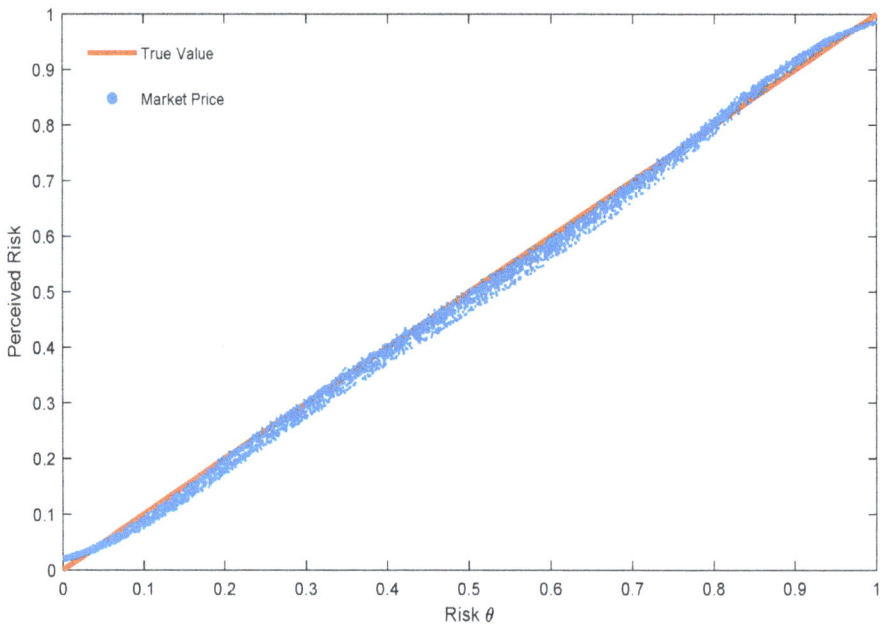

Figure 10.7: Price $\phi_L(\mathbf{z}\boldsymbol{\mu})$ versus θ for last 5,000 plays of 100,000 given $\mathbf{z} \equiv [(2\theta-1)^i]_{i=0}^7$, $b_{ik}(t) = 10^{-j}(t+5)^{-k/4}$ for $j,k = 0,\ldots,4$, and $\lambda = 10^{-6}$

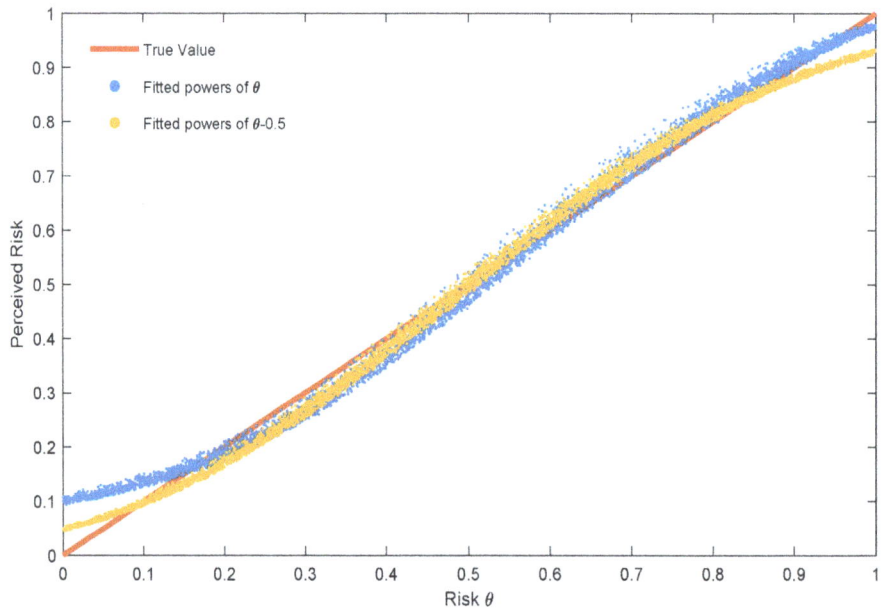

Figure 10.8: Analogues to Figure 10.7 when $\mathbf{z} \equiv [\theta^i]_{i=0}^7$ or $\mathbf{z} \equiv [(\theta-0.5)^i]_{i=0}^7$

11

Futures Forecasts

The more deferred the outcome, the less the next observation can tell us about it. If we aim to forecast the climate a million years from now, current weather offers no substantive insight compared to new research perspectives on the past. All a market on that climate can tell us is the evolving consensus. Nevertheless, tomorrow's weather might slightly shift the credibility of various notions and thereby modify the consensus.

To explore this, let's imagine a gambling table at Harold's Casino that trades a present claim on the next Bernoulli x and a future claim on the x 10,000 periods from now. As before, suppose θ resets at rate $\Lambda = 0.001$ from a uniform distribution. Since the chance of no resampling in 10,000 periods is less than 0.005%, the true value $\left\langle \theta^{fut} \right\rangle$ of the future is very close to 0.5 regardless of the current θ. However, the gamblers don't know the true reset parameters. Instead, they anticipate reset with probability λ from a beta (α, β) distribution, which can have any mean and variance that the range of x allows and includes uniform as the special case $\alpha = \beta = 1$. I call this a Beta Reset (BR) belief.

For a Perfect Learner, the RMSE for the present claim is 0.045. Divide λ by 10 or triple α and the RMSE increases by 8%. Even for $(\alpha, \beta, \lambda) = (5, 0.5, 10^{-5})$, which overstates the long-term mean by 80% and understates Λ by factor of 100, the RMSE for the present claim is 0.061, just one-third above the minimum.

In contrast, forecasts of the future claim are highly sensitive to the BR belief. For a Perfect Learner, the RMSE for the future claim is just two parts in a million. Divide λ by 10 and the RMSE jumps to 0.107. Divide λ by 100 and the RMSE reaches 0.26, which makes the forecast barely more illuminating than a random guess. When λ exceeds Λ, the RMSE is $|\mu - 0.5|$; e.g., 0.167 for $(\alpha, \beta) = (2,1)$ and 0.25 for $(\alpha, \beta) = (3,1)$.

To improve the forecasts, I set a base layer of 512 different BR predictors with three layers of TDs above. The base layer offers 8 choices for each parameter, with α and β spaced geometrically between 0.2 and 5 and λ spaced geometrically between 10^{-1} and 10^{-6}. The first two layers of TDs consist of 8 bots each with doubts of 10^{-i} for $i = 1, \ldots, 8$. The top layer of TDs applies the median doubt. I simulated risks and outcomes for 100,000 plays, starting with beta-distributed priors on θ for each BR predictor and uniform priors for the various TD layers. After every outcome I updated beliefs as if only the present claim were traded. The future claim I priced as follows:

- For each predictor k, the fair price E_k^{fut} at future time T is a weighted average of the BR mean and the present fair price E_k, with the latter weighted by the probability $(1 - \lambda_k)^{-T} \approx \exp(-\lambda_k T)$ that θ doesn't change in the interim.
- The consensus future price E^{fut} equals the credibility-weighted average of the E_k^{fut}.
- Credibility adjusts in line with wealth from short-term trading.

Figure 11.1 displays sample paths over 20,000 observations for E and E^{fut}. Clearly, E tracks its target far better than E^{fut} does. The latter gets pulled too much toward current E, although the pull appears to diminish over time.

Figure 11.2 displays RMSEs out to time $t = 100,000$ with each measure computed across 200 simulations and all observations within 100 periods of the specified time t. The RMSE for E^{fut} halves over the first 10,000 observations, during which θ switches 10 times on average. After 40,000 observations, E^{fut} tracks its underlying value better than E does. Long-term tracking continues to improve slowly thereafter. However, even after

Figure 11.1: Sample paths to time $t = 20{,}000$ in a Beta Reset market for present claim, 10,000-period-future claim and true risks

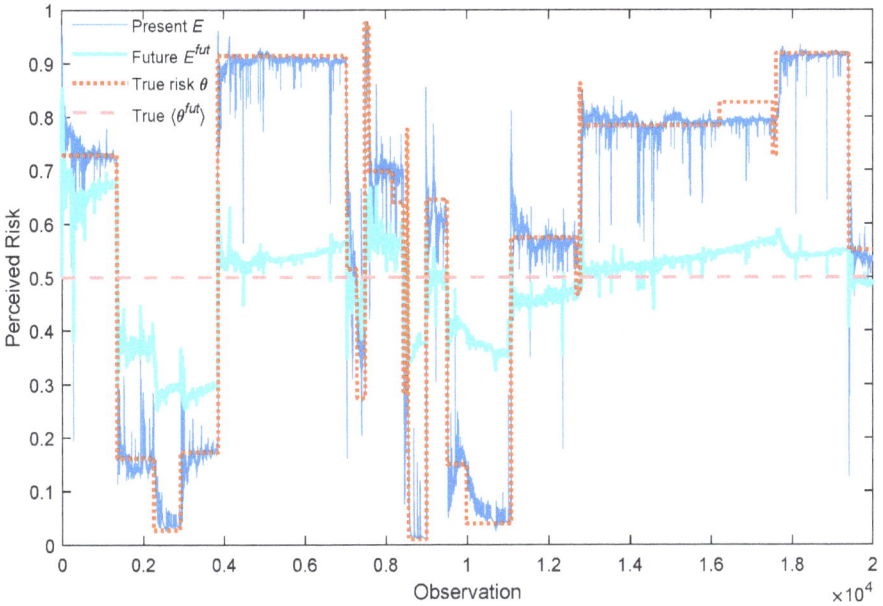

Figure 11.2: RMSE for E and E^{fut} to $t = 10^5$ computed over 200 simulations and all observations within 100 periods of t

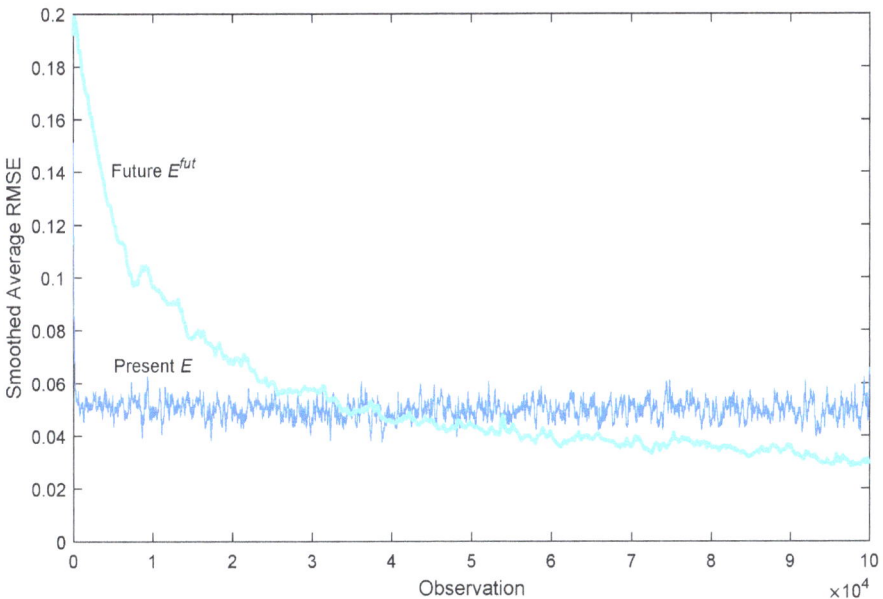

100,000 observations or 100 switches in θ, E^{fut} often diverges from the nearly constant $\langle \theta^{fut} \rangle$ by 0.02 or more.

Suppose we try to improve E^{fut} tracking by focusing on 10,000-day outcomes. In 100,000 days, we obtain only 10 independent observations. While we can augment these with nearly 90,000 overlapping 10,000-day observations, the latter don't improve the RMSEs very much. They are too highly correlated with what they overlap. Bear in mind that the evidence is generated by only about 100 different θ. Even if measured perfectly, they won't reveal the mean of the distribution. For random draws from a uniform distribution, the standard error of the empirical mean is 0.029. That's not much better than the Beta Reset market achieves.

Adding to the problems, Bayesian updates that rotate among bad long-term predictors don't always generate good long-term predictors. For example, suppose we replace the base-level BRs with "EMA Reverters", which nudge the prediction toward the current observation and then let it revert toward an assumed long-term mean. I let the long-term mean, the short-term nudge, and the reversion speed take eight values each, with the first spaced linearly on the unit interval, the second spaced geometrically between 0.3 and 0.003, and the third spaced geometrically between 0.001 and 0.0001. Three layers of TDs sit above, with the same parameters as in the previous example.

Figure 11.3 displays sample pricing paths over 20,000 observations for the EMA Reverter ensemble. The fit for E isn't bad. While the path wobbles more than in Figure 11.1, it has fewer spikes. However, E^{fut} doesn't fit 0.5 as well as a simple average of 0.5 and E would.

In this case, futures prediction is so bad that it begs an ad hoc patch. Suppose we estimate E^{fut} as the simple average of all observed x. Figure 11.4 plots the smoothed RMSE for 200 simulations, superimposed on the RMSEs for E^{fut} from Figures 11.2 and 11.3. The trajectory mostly parallels that for the BR market, with the excess RMSE shrinking to about 0.01 after 100,000 observations. Eventually the simple average will so outperform the E^{fut} predictions of EMA Reverts that it will force a reappraisal. Models that let θ wander short-term but keep the long-term mean close to 0.5 will be favored, encouraging evolution toward a BR-type solution.

Figure 11.3: Sample paths to time $t = 20,000$ for E, E^{fut}, and θ when EMA Reverters replacing Beta Resets

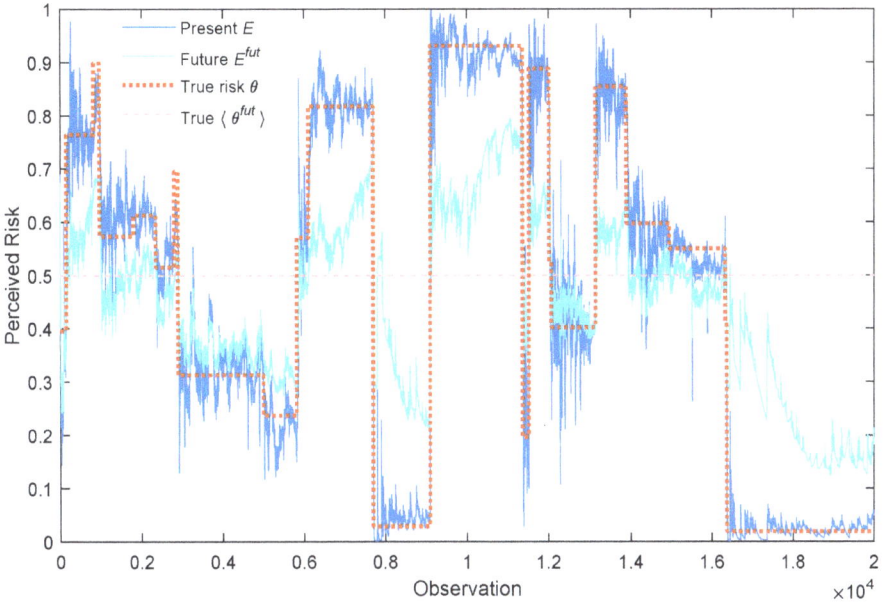

Figure 11.4: RMSEs for E^{fut} comparing historical average returns with estimates from Beta Reset and EMA Reverter markets

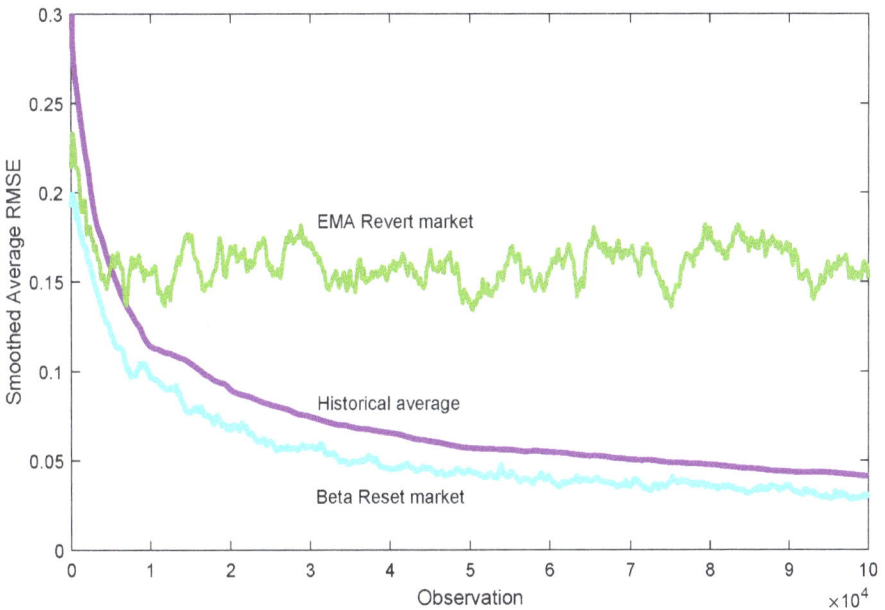

Futures Markets

Each observation x informs about trajectories at multiple time horizons. For Perfect Learners, the longer trajectories offer no extra insight. For learners who don't know how the risk was generated, observations on the longer trajectories help filter out inappropriate models. Markets in future outcomes, aka futures, potentially assist the filtering by boosting gamblers who predict the long-term well. However, the connections are subtler and more tenuous than for short-term markets.

Note that any claim on future outcomes can be purchased next period instead, with more information both on likelihoods and wealth. Current purchase is merited only if the gambler anticipates taking advantage of a quick shift in market consensus about the future. Suppose the market assigns a fair price of E for today's outcome x_o and $E + \varepsilon$ for tomorrow's outcome x_1, with interest and inflation ignored and with θ presumed fixed for the day. If $\varepsilon = 0$, tomorrow's claim will be valued tonight at either V/E more or $V/(1-E)$ less than it was this morning. Since the same returns can be generated by buying $\frac{V}{E(1-E)} < 1$ units of today's claims, the futures market essentially replicates the one-period market.

Next let's suppose $\varepsilon \neq 0$. If the impact is independent of x_0, the price tonight of tomorrow's claim will again be either V/E more or $V/(1-E)$ less than it was this morning. This leaves profits or losses the same as before and the futures market just as redundant. If the impact varies with x but the initial price of $E + \varepsilon$ is a fair market valuation, it must be possible to express tonight's anticipated price for tomorrow's claim as $E + \varepsilon + V_{alt}/E$ or $E + \varepsilon - V_{alt}/(1-E)$ for some alternative V_{alt}. This generates net returns of V_{alt}/E or $-V_{alt}/(1-E)$, which again just rescales the one-period market.

The main justification for preferring the futures contract to a present contract, apart from potentially easier settlement,[5] is anticipation of a one-period shift in ε. Yet ε itself represents an anticipated shift in the underlying risk θ. Hence, heavy trading in long-deferred futures rests on anticipation of a shift in the anticipation of a shift. That seems strange. By construction, beliefs about the current θ affect E only, not ε. Observation of x_0 shouldn't change ε either.

However, x_0 isn't all that is observed as time passes. Time offers opportunities to reappraise history. Perhaps we'll come up with new hypotheses we hadn't thought of before or reevaluate old hypotheses for reasons unconnected with current evidence x_0. Call it a change in paradigm or conceptual framework. That can shift ε. We have already discussed a stylized example inspired by Figure 11.4, where short-term E is initially forecast using EMA Reverts only but the latter are eventually discredited by poor prediction of E^{fut}.

Granted, major reappraisals are sporadic. How can anticipation of them sustain a futures market? The answer is that paradigms don't change quickly or smoothly. It takes time to refine hypotheses, to re-weigh the evidence, and to persuade others. Betting on convergence might justify a long-term hold as that hedges out uncertainty over timing.

Since a two-period model can't show that, let's redo the analysis with a 10,000-day future purchased today for E^{fut}. Tomorrow it can be resold at the market price for a 9,999-period future. Suppose gambler i estimates the profit as

$$\delta_i \left(x_0 - E \right) + \Delta \varepsilon_i \, ,$$

where both δ_i and $\Delta \varepsilon_i$ are independent of x_0. Let's reflect on the relative influences of these parameters. For $\Delta \varepsilon_i$ a potential mispricing of E^{fut} on the order of 0.1 coupled with uniform chances of correction over the next 10,000 days imply an order of magnitude of 10^{-5}. For δ_i, competition should drive that towards the Perfect Learner $\delta_{PL} = e^{-10000\Lambda} \cdot V / E(1-E)$, which in our simulations is typically on the order of 10^{-6}. Moreover, the expected impact of the δ_i term is diluted by the factor $E_i - E$ and can be hedged in the market for x_0. This confirms that active futures trading is driven far more by perceptions of δ_i than by perceptions of $\Delta \varepsilon_i$.

A gambler i who perceives variance σ_i^2 for the hedged future and applies Kelly fraction f_i will bet $f_i (\Delta \varepsilon_i - \Delta \varepsilon)/\sigma_i^2$ given consensus shift $\Delta \varepsilon$ with equilibrium

$$\Delta \varepsilon^* = \sum_i \frac{\kappa_i f_i}{\sigma_i^2} \Delta \varepsilon_i \, \Big/ \sum_i \frac{\kappa_i f_i}{\sigma_i^2} \, .$$

Nominally this looks very similar to the market for the present claim, with $\Delta\varepsilon_i$ replacing E and σ_i^2 replacing the conditional variance $E(1-E)$. However, there's a huge difference in substance. Since tomorrow's net return hinges on a shift in subjective consensus, strategic gambling in futures markets requires forecasts of others' beliefs.

For example, suppose that E^{fut} at Harold's Casino is currently priced at 0.45. While we are confident it is worth 0.5, a rumor is spreading that Harold wants to keep average θ below 0.5. We expect the rumor to gain credence if θ stays low, which happens to be our near-term forecast. On reflection, we should defer some purchase of E^{fut} or sell some nearer-term futures. Furthermore, the credence the rumor has already gained might challenge our conviction. Might others know something about Harold that we don't? Perhaps we should trim our Kelly fraction f.

Bear in mind too that consensus can be partially self-fulfilling. Market optimism about a firm's prospects can raise its capital value, which in turn facilitates investments that can justify its value. This is yet another factor encouraging participants to defer to consensus or to bet on the momentum of perceived trends in consensus.

In principle, a well-developed futures market, including options on combinations of outcomes, allows identification of market beliefs in fine detail. Yet that doesn't make it easy to hedge them and trade solely on fundamentals. Who is likely to profit most from futures trading? Not the deep analyst of beauty. Not the slave to consensus. The profits flow best to those just ahead of the crowd and watching where it is leaning. These conclusions echo Keynes (1936, Chapter 12):

> The energies and skill of the professional investor and speculator are mainly occupied [...] not with making superior long-term forecasts of the probable yield of an investment over its whole life, but with foreseeing changes in the conventional basis of valuation a short time ahead of the general public.... Moreover, this behavior is not the outcome of a wrong-headed propensity. It is an inevitable result of an investment market organized along the lines described.

PART III: CREDIT MARKETS

Credit markets are like casino games in which payment is standard and default the rare exception. The asymmetry makes credit spreads jump on bad news and gradually shrink on good news. High safety is relatively uncertain, as it is hard to distinguish the risk of credits that hardly ever default. Anticipation of possible change in risk makes debt markets look myopic. Markets in sovereign debt appear both uncertain and myopic.

Credit ratings are noisy indicators of a relative default risk that changes over time. Like aggregate risk, relative risk is hardest to identify among the seemingly best credits. Reputable long-duration government debt is especially vulnerable to surprise. International banking regulation focuses too much on reducing current default risks on banks' assets and too little on limiting duration mismatch.

Rational myopia can make lenders too complacent in booms and too pessimistic in crisis. It lets sovereigns with a history of good servicing borrow too much for too long and then slams them with the repercussions. On the bright side, it helps speed restoration of confidence after a crisis.

12

Default Risk

From a modeling perspective, debt markets are casino games in which one outcome is far more likely than the rest. That outcome, called servicing, pays what was promised. Any failure, called default, disrupts all future payments. I will assume servicing is continuous until default and that default is followed immediately by issuance of similar new debt. I will call this irregular cycle a credit. Let v denote the instantaneous default risk so that the default risk θ over a short interval dt is approximately $v\,dt$. If p denotes the perceived probability density of v with mean E, Bayesian updates require

$$
\begin{aligned}
p(v) &\to \frac{p(v)v}{E} && \text{on default} \\
\frac{dp(v)}{dt} &= p(v)\bigl(E-v\bigr) && \text{otherwise}
\end{aligned}
\tag{12.1}
$$

to which we can add adjustment for doubts and anticipated change. The asymmetry between jump on news of default and gradual shrinkage on news of servicing is the most outstanding feature of credit markets.

Forecasters often opt for a simple counting routine, where they record D_{eff} effective defaults in T_{eff} effective time and estimate mean default risk E as their ratio. We can reconcile this with (12.1) by defining $T_{eff} \equiv E/V$ and $D_{eff} \equiv E^2/V$, in which case

$$\Delta E = \frac{V}{E} = \frac{1}{T_{eff}} = \frac{E}{D_{eff}} \quad \text{on default}$$
$$\frac{dE}{dt} = -V = -\frac{E}{T_{eff}} \quad \text{otherwise}$$

$$(12.2)$$

In the standard counting routine, D_{eff} jumps by one at every default and otherwise stays the same, while T_{eff} adds one every observation. This is rational provided risk never changes and p is gamma-distributed with shape parameter D_{eff} and inverse scale parameter T_{eff}. At unit shape, a gamma density is exponential, with decay rate $1/T_{eff}$. For $D_{eff} \gg 1$, gamma densities look Gaussian, except that they prohibit negative v and skew slightly positive. For $D_{eff} \ll 1$, gamma densities look L-shaped, with infinite steepness at the origin and a long, relatively flat tail. Figure 12.1 charts gamma densities for various shapes and common mean.

Small fractional shapes are needed to allow for vague prior beliefs. $D_{eff} < 1$ indicates that the standard deviation $\sqrt{V}$ of beliefs exceeds the mean E, in which case E will more than double after a relevant default. There's no reason to rule this out. Even though each default propels D_{eff} over one, both D_{eff} and T_{eff} shrink when there is no default and risk is perceived to change. In most simulations of highly trusted debt, D_{eff} has to be tiny. For example, consensus E for Japanese government bonds is typically a few bps a year. Yet it is hard to identify even a thousand years of relevant independent data on Japan's current default risks. If $T_{eff} = 300$ years, any $D_{eff} > 0.3$ implies $E > 10$ bps per year.

While no uncertainties about default are exactly gamma-distributed, the dominant shape tends to be gamma-like. In particular, D_{eff} can't be less than 1 without concentrating most of the density well below the mean. Since default crushes beliefs closest to zero, it can multiply E manyfold without implying complacency before or over-reaction after. The next few chapters will show that low D_{eff} is quite common in credit markets and helps explain their characteristic features. More precisely, I demonstrate the plausibility and appeal of such explanations, through simulations that look broadly consistent with the historical evidence.

Figure 12.1: Gamma densities for various shapes with means held constant.

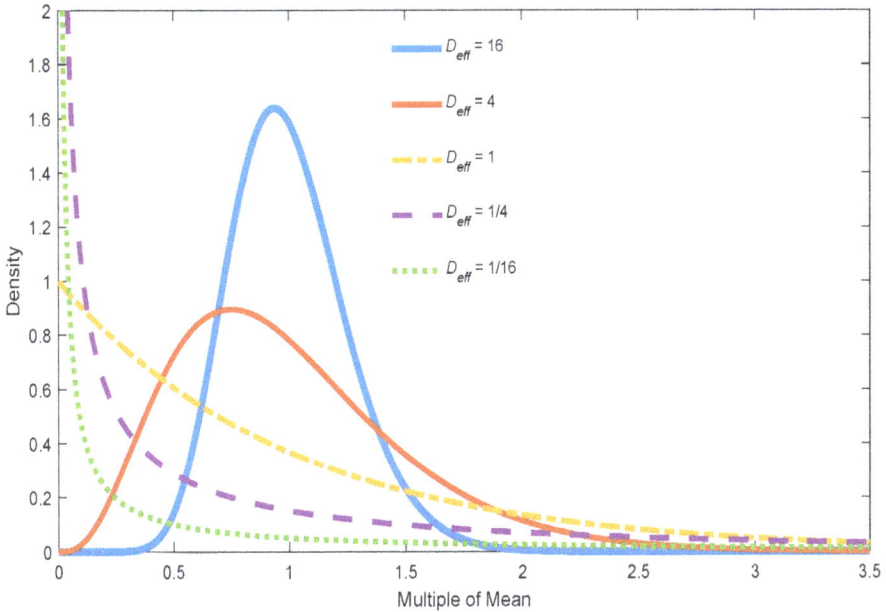

On a first pass let's ignore variations across debtors and focus on average risk v in any given year. I will calibrate one set of risk parameters for corporate debt, where the data is relatively abundant and easy to work with, and a different set for sovereign (government) debt.

Figure 12.2 cobbles together various sources to display annual corporate default rates for the past 150 years. They are measured in basis points (bps) per year, where 100 bps equals one percent. The chart is inspired by Kay Giesecke et al (2010) and largely replicates their Figure 1. The rates are volume-weighted until 1920 and an average of volume-weighted and issuer-weighted later.[1] The data cover the US only for the first 50 years and gradually expand abroad; in recent years roughly a third of the data comes from outside the US. The number of issuers fluctuated between 500 and 2500 through most of the period but has surged in recent decades: S&P rated 6900 issuers in 2017, over triple the number 25 years prior.

The data from the late 19th century likely deserve less weight than the rest. US corporations were just getting their footing then, with worse management than is typical today. Macroeconomic management was worse

Figure 12.2: Annual US/global corporate default rates 1866-2017.

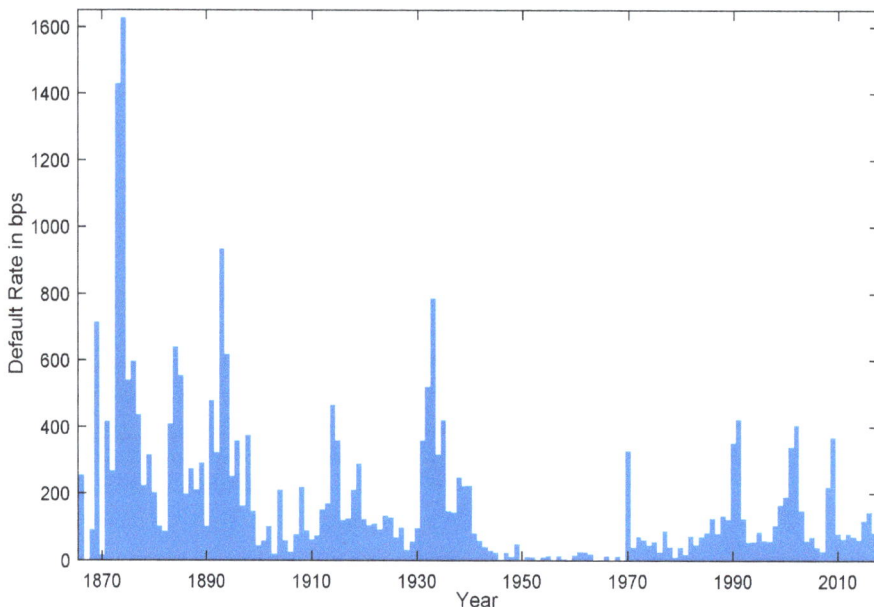

too, making the crash in 1873 more severe than any peacetime crash since. Also, the credit pool was much smaller and weighting credits by value heightens their effective concentration. Yet dropping the 19th century data would exacerbate other biases. The 20th century data focus on the main victor of two intense world wars and one protracted cold war. The 21st century data don't yet include a reckoning for the massive buildup of corporate leverage. Furthermore, post-1920 data covers only corporates with formal credit ratings, which biases the averages downward.

The extended fluctuations and high dispersion make it hard to identify an equilibrium distribution. For the data underlying Figure 12.2, the sample mean E_0 is 177 bps and the best-fitting gamma shape D_0 assuming risk changes only at year end is 0.80.[2] To model the evolution of risk, I invoke the simplest diffusion process that can generate a long-term gamma distribution with those parameters. It presumes that $\log v$ has constant volatility s, in which case its drift must decrease linearly with v. Replacing Δ with the differential d and letting dz denote standard Brownian motion, the infinitesimal changes in $\log v$ satisfy

$$d \log v = \tfrac{1}{2} s^2 D_0 \left(1 - V\!\!\left/\!E_0 \right. \right) dt + s\,dz \,. \qquad (12.3)^3$$

The data exhibit a median 0.61 absolute change in yearly log default rates and an average 10-year cycle of yearly default rates surging above the mean and falling below. Either metric is fit best by $s \approx 1.2$.[4] At that volatility, the smoothing caused by yearly averaging of v exaggerates apparent shape by nearly a third. Rounding the best fits, my corporate benchmark sets $E_0 = 180$ bps, $D_0 = 0.6$ and $s = 1.2$. However, these values don't warrant high confidence. In repeated simulations, the standard deviations exceed 35 bps for E_0, 0.2 for D_0, and 0.25 for s. Nor can we be sure that (12.3) fully describes the drivers of v or that the parameters are stable.

My corporate benchmark also sets pool size $N = 2500$. The choice is a compromise between the larger values of recent years, the mostly smaller values before 1990, and the effective compression wrought by value-weighting and by fluctuations in the number of issuers.[5] It roughly matches the average number of corporates tracked each year by rating agencies since 1920.

Turning to sovereign debt, the best collection of data comes from spreadsheets maintained online by Carmen Reinhart, following up on Reinhart and Rogoff (2009). Their core measures identify the entire interval between default and resolution as the time **in** default. To convert these into default rates I treated the first year of each default episode as the time **of** default and the rest of the episode as a limbo not registering either default or servicing. In other words, I computed each year's default rate as the number of default episodes starting that year divided by the number of issuers who weren't in default a year prior.

Figure 12.3 depicts sovereign default rates for 1800 through 2012. Their average is approximately 270 bps per annum, 50% higher than the average corporate rate in Figure 12.2. The annual rates jump around far more than corporate rates do, mostly because the pools are much smaller. Figure 12.3 covers only 17 independent countries in 1800, 44 in 1900 and 125 in 2012. The average active pool size per year has been 51 with an additional 11 sovereigns in limbo.

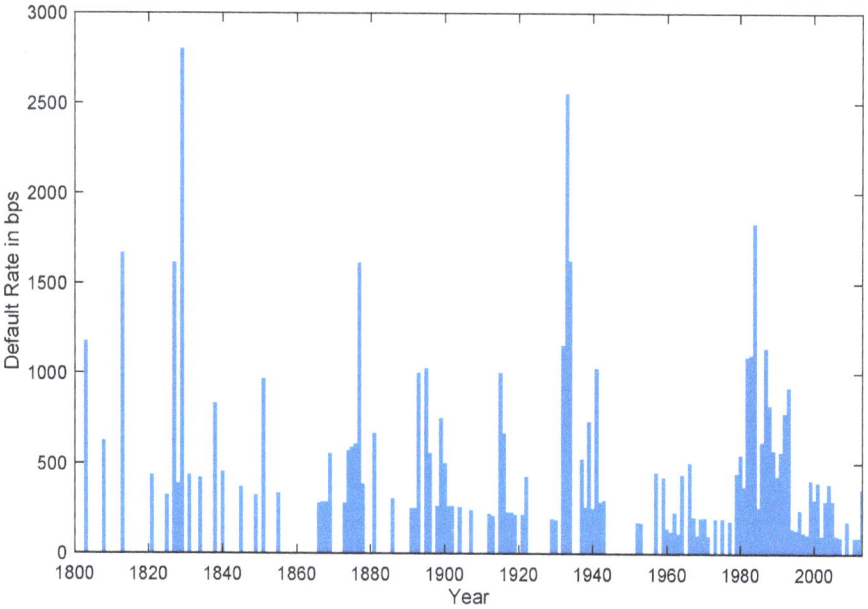

Figure 12.3: Sovereign default rates on external debt 1800-2012

Many of the newcomers are small countries with implicit guarantees from larger countries, reducing their statistical independence. However, some of the extra jumpiness relative to Figure 12.2 appears to reflect speedier diffusion. Simulations of (12.3) fit the data best when $D_0 \approx 0.6$ and $s \approx 1.7$.[6] Again, that leaves considerable uncertainty. The standard deviations exceed 40 bps for E_0, 0.15 for D_0 and 0.4 for s.

I choose $\Lambda = 0.02$ for the sovereign benchmark to remind of the inherent fragility of global governance. This implies one reset on average every fifty years, which is far less disruptive than it might appear. Half of resets lower risk. On average, only one in five resets will hike v by more than E_0 and only one in ten resets will hike v by more than $2E_0$. For sovereign pool size I choose $N = 80$, which roughly matches the yearly average since 1920. To summarize, the benchmarks are:

- Corporate: $E_0 = 180$ bps, $D_0 = 0.6$, $s = 1.2$, $\Lambda = 0$, $N = 2500$.
- Sovereign: $E_0 = 270$ bps, $D_0 = 0.6$, $s = 1.7$, $\Lambda = 0.02$, $N = 80$.

Simulations

When a pool of N iid credits share a common v, learning speeds up by a factor of N. While E still jumps by $1/T_{\text{eff}}$ when default occurs, N times as many defaults are expected in a given time. When no default occurs, E shrinks N times as fast. This modifies (12.1) to

$$
\begin{aligned}
p(v) &\to \frac{p(v)v}{E} && \text{on default} \\
\frac{dp(v)}{dt} &= Np(v)(E-v) && \text{otherwise}
\end{aligned}
, \qquad (12.4)^7
$$

The simulations below model the evolution of risks, defaults, and beliefs 250 trading days per year for 100,000 years. Credits that randomly default are replaced by others, keeping N constant. The market is treated as a Perfect Learner, who knows all the evolutionary parameters.

For the corporate benchmark, E tracks v well with a correlation of 0.93, an RMSE of 90 bps and a median absolute deviation of 26 bps. Figure 12.4(a) plots 100 years of daily tracking. The transitions between calm and turbulent phases largely mirror the changes in v. However, (12.4) induces a striking asymmetry, where E soars on default and retreats gradually on improved servicing. Comparing average annual values, the correlation between E and v rises to 0.98 and the RMSE shrinks to 40 bps. Half of the absolute annual deviations are less than 15 bps and only 3% exceed 100 bps. Figure 12.4(b), which plots average annual E and v over 500 simulated years, illustrates the generally close correspondence.

For the sovereign benchmark, tracking is much worse. Correlation between E and v drops to 0.64. The RMSE and median absolute deviation triple or quadruple to 268 bps and 105 bps respectively, whereas higher sovereign E_0 explains at most a 50% rise. Furthermore, sovereign E_0 fails to capture extreme values of v. It never reaches 2500 bps and never falls below 24 bps, whereas v can reach 4500 bps and falls below 24 bps 19% of the time.

Figure 12.4: Simulated tracking of default risk for corporate benchmark

(a) Daily tracking for 100 years

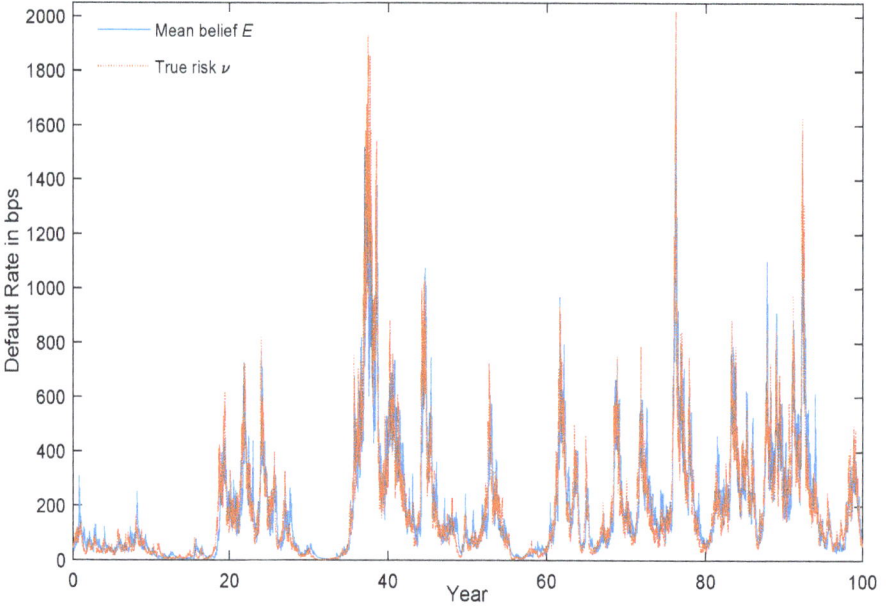

(b) Average annual values for 500 years

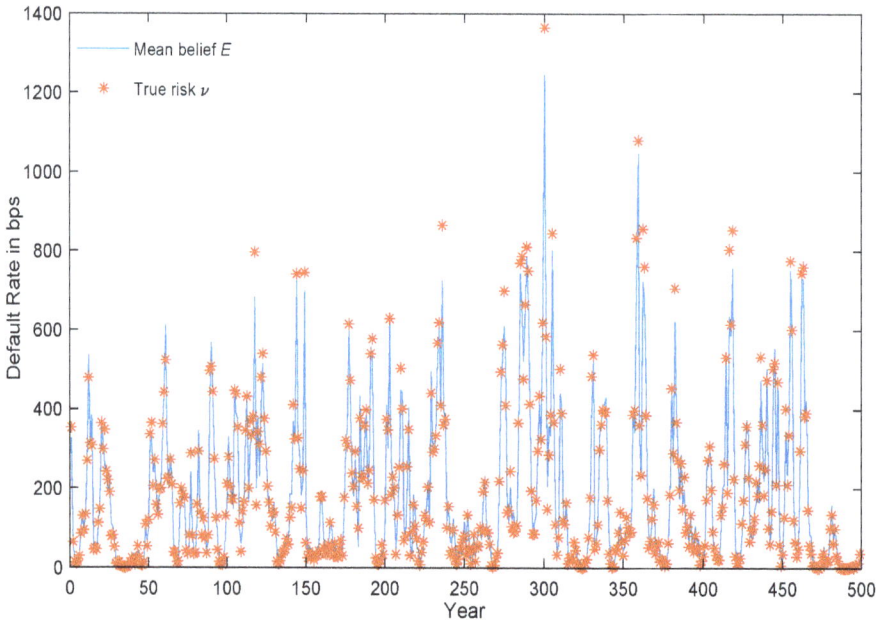

Figure 12.5(a) plots daily tracking for 100 years. Adjustments to lower risk noticeably lag, with some significant deviations persisting for many years. Comparing average annual values, the correlation between E and v is 0.79, the RMSE is175 bps, the median absolute deviation is 90 bps and 3% of deviations exceed 400 bps. Figure 12.5(b) highlights the frequent discordance between average annual E and v over 500 years.

Figure 12.5 depicts totally rational behavior, in which E tracks v with some lag and smoothing. In practice, no one sees v, which can make E look unduly sensitive to recent default rates. Dividing the effective sample size T_{eff} by pool size N measures the effective sample duration. What is a reasonable value? Here it never reaches a year. Mean duration is three months for the corporate benchmark and six months for the sovereign benchmark. That sounds terribly myopic. However, the Perfect Learner is just balancing the marginal imprecision from underweighting evidence on similar risks against the pollution from overweighting evidence on dissimilar risks. To say this informally, it would rather be approximately right than precisely wrong.

The market will look most myopic when default occurs at low E. Figure 12.6 displays scatter plots of D_{eff} versus E for both benchmarks. Plotted on a log-log scale, they resemble thin upward-sloping feathers with narrow quills at the bottom. For the corporate benchmark, no $E<8$ bps exhibits $D_{eff}>1$, no $E<25$ bps exhibits $D_{eff}>2$ and no $E<83$ bps exhibits $D_{eff}>4$. For the sovereign benchmark, no $E<235$ bps exhibits $D_{eff}>1$ and no $E<520$ bps exhibits $D_{eff}>2$.

For perspective, consider ratings of pharmaceutical safety. Suppose a new medicine has been tested on T patients without a single instance of harm. How safe is it? The chance of no harm in T iid trials is 14% for a true risk of $2/T$, 5% for a true risk of $3/T$ and 2% for a true risk of $3.9/T$. As medical regulators are extremely cautious, they don't want to rule out any of these possibilities. Accordingly, T trials without harm is typically viewed as justifying at best a claim that risk is less than $4/T$. If we use $4/T$ as an analogously cautious bound on v, our corporate benchmark would never justify high confidence that v is less than 20 bps while our sovereign benchmark would never justify high confidence that v is less than 500 bps.

Figure 12.5: Simulated tracking of default risk for sovereign benchmark

(a) Daily tracking for 100 years

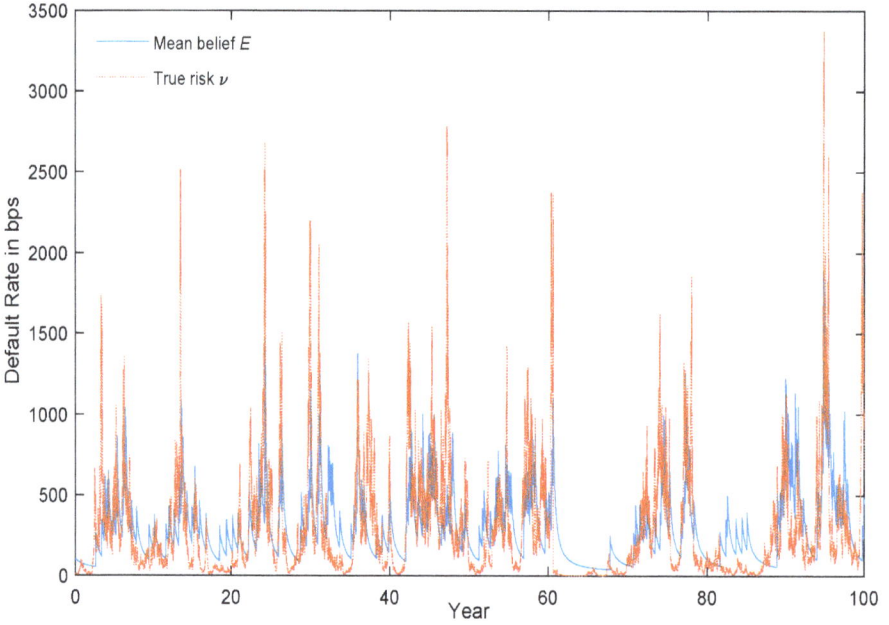

(b) Average annual values for 500 years

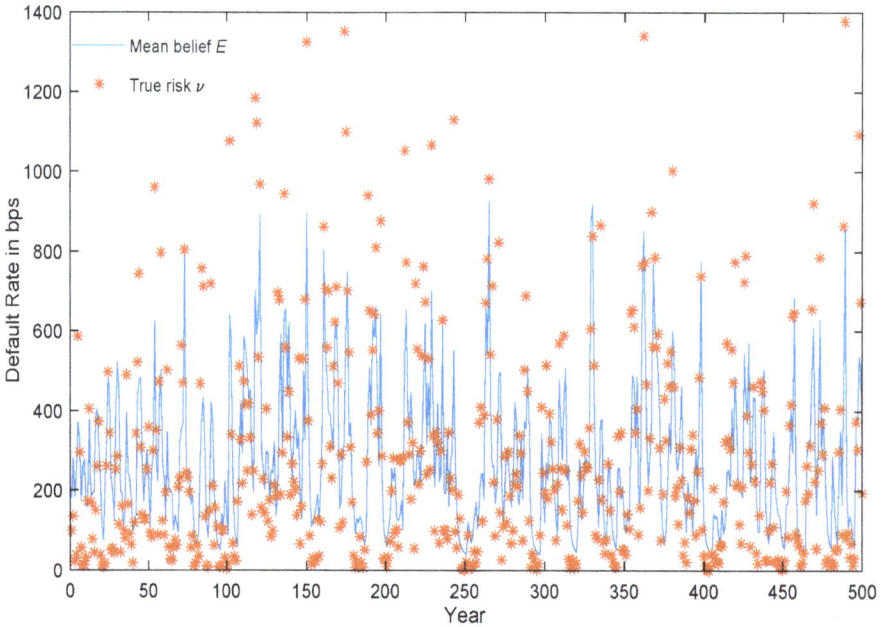

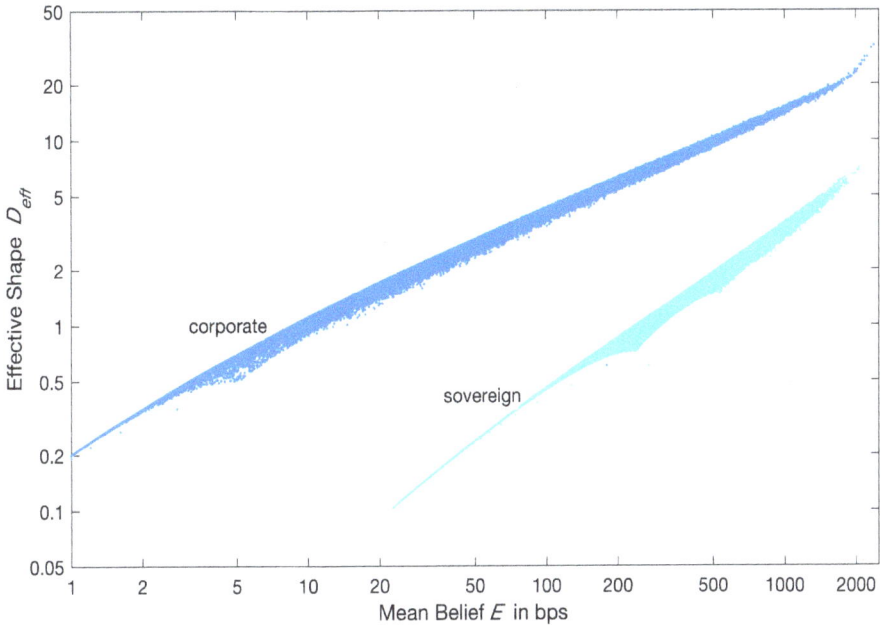

Figure 12.6: Scatter plot of D_{eff} versus E for corporate and sovereign benchmarks

This does not mean that low E tends to understate the true risk v. I am just emphasizing the distinction between expectation and assurance. Safe debt is like weather in the Caribbean: very mild except for the hurricanes. Owning such debt is like insuring against hurricanes, both in the rarity of claims and in the uncertainty over what the current risks really are.

Behaviorists emphasize human propensities to get complacent about disaster risks and to overreact when disaster strikes. Yet our simulations show purely rational bots pricing risk similarly. High safety is always relatively uncertain, due to the difficulty in gauging the rarity of an exception. When default doesn't occur, bots aren't sure how long they should expect to wait until it does. When default does occur, they're tempted to conclude that it's not so rare after all.

13

Bond Pricing

Most borrowing on credit markets is done for extended periods. A classic bond pays interest periodically but no principal until the maturity T. Loans typically carry cross-default protection that triggers default on all obligations if any required payment is missed. Defaulting borrowers are typically shut out of credit markets until they provide some compensation and offer new protections. This gives defaulted debt some salvage value. Default risk v will vary with the volume of debt outstanding, the ease of refinancing, the need for additional loans, their interest rates, and the anticipated penalties for default. Not all these will be clear, so the market's beliefs about v will influence its actual value.

To skirt around these complexities. I will focus on bullet bonds, which pay nothing until maturity. I will assume that v evolves exogenously, uninfluenced by trader beliefs. I will ignore impatience, inflation, and risk aversion. In this chapter I will ignore salvage value too. This simplifies fair price per unit principal to the survival probability of a random process, where default triggers extinction. My simulations set a benchmark T of 10 years, which makes most bullets behave like classic bonds of 11-14 years maturity.

Suppose we know the survival probability $S_t(v_j)$ to time t for any initial default risk v_j. Then $S_{t+\Delta t}(v_i)$ equals the survival rate $\exp(-v_i \Delta t)$ times a weighted average of the $S_t(v_j)$, with weights given by the transition probabilities from v_i to the various v_j. This lets us compute S_t by recursion starting from $S_0 \equiv 1$.[8]

Figure 13.1 plots S_{10} versus v for the corporate and sovereign benchmarks. The recursion tracks S_t 250 days a year for 10 years over a grid of 500 different v.[9] The sovereign curve lies below the corporate curve due to higher long-term average default risk. Both curves are highly convex at low v, which has an interesting implication. By Jensen's inequality, the survival probability will be higher the more uncertain the beliefs about v are, holding mean belief E constant.

In credit markets, prices are commonly quoted in terms of implied credit spreads, which indicate the effective interest rate premium over the risk-free rate. Under our assumptions, the instantaneous credit spread should equal the instantaneous default risk v. When v stays constant, a fraction e^{-vT} of credits are expected to service for at least T years. Hence, a bullet's expected survival rate of S_T is associated with a credit spread of $c_T \equiv -\log S_T / T$.

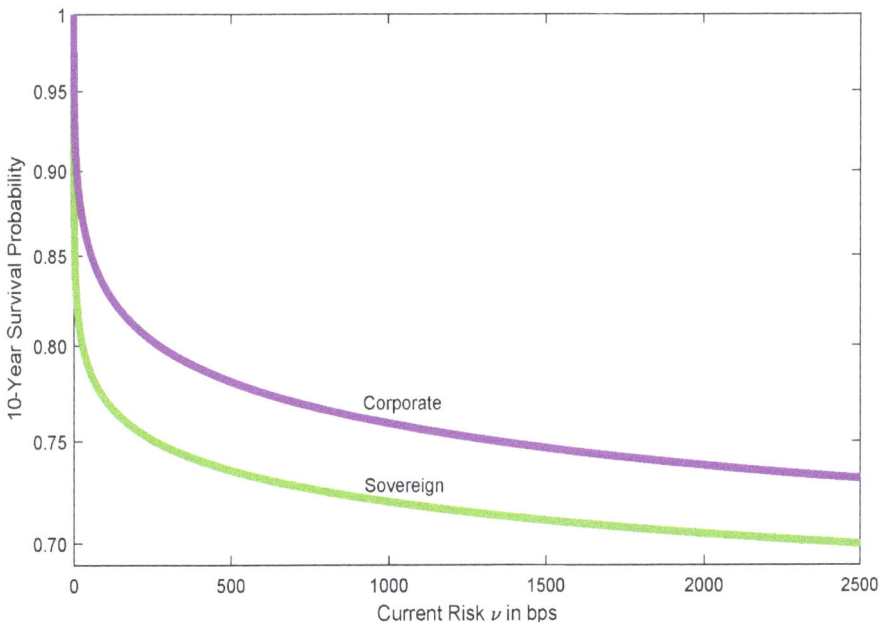

Figure 13.1: 10-year survival probabilities for corporate and sovereign benchmarks

Figure 13.2 recasts Figure 13.1 in terms of c_{10}, with log spacing on the horizontal axis to better display the values for low v. The range of c_{10} is on the order of 250 bps, far narrower than the 2000+ bps range of v. This indicates that diffusion is expected to rapidly reduce the impact of the starting point. Doubling c_{10} requires on the order of a tenfold increase in initial corporate v or a hundredfold increase in initial sovereign v. The sovereign curve is flatter than the corporate curve because sovereign risk diffuses faster.

Due to convexity, the fair credit spread $C_T \equiv -\log\langle S_T \rangle / T$ for uncertain beliefs with mean E is less than the spread c_T for diffusions that start at $v(0) = E$, while c_T is less than the expected default rate $c_T^* \equiv \int_0^T \langle v(t) \rangle dt / T$. Figure 13.3 displays the comparisons from a 100,000-year simulation. For the corporate benchmark, the total $c_T^* - C_T$ gap widens gradually from 8 bps at $E = 2000$ bps to 18 bps at $E = 0.5$ bp. For the sovereign benchmark, $c_T^* - C_T$ widens much more, growing from 12 bps at $E = 2000$ bps to 34 bps at $E = 100$ bps and 73 bps at $E = 25$ bps.

Note that the sovereign scatter plot doesn't extend nearly as widely as the others. Between anticipated reversion and limited observation, a Perfect Learner never gets fully persuaded that v is extremely low or high. Its perfection refers only to its knowledge of the underlying process and its rational inference from observation. Neither rule out bias at the extremes.

Bayesian bias sounds like an oxymoron since rational analysts should strive to correct it. However, they don't know enough to do so. The impact for sovereigns is asymmetric. Bayesian bias helps offset the understatement of current v or E by low credit spreads but aggravates the understatement by high credit spreads. The projected sovereign $C_{10} = 155$ bps for $E = 25$ bps is close to the mean 10-year sovereign risk for $v = 1$ bp. Yet E never reaches 2400 bps and C_{10} never reaches 355 bps even though v in simulations occasionally exceeds 4000 bps.

According to Reinhart and Rogoff (2009), lenders to sovereigns tend to understate big default risks. In our simulations, so do perfectly rational learners. This makes Reinhart and Rogoff's findings more plausible but challenges their inference of lender irrationality. While they surveyed all

Figure 13.2: Implied 10-year credit spreads c_{10} for Figure 13.1

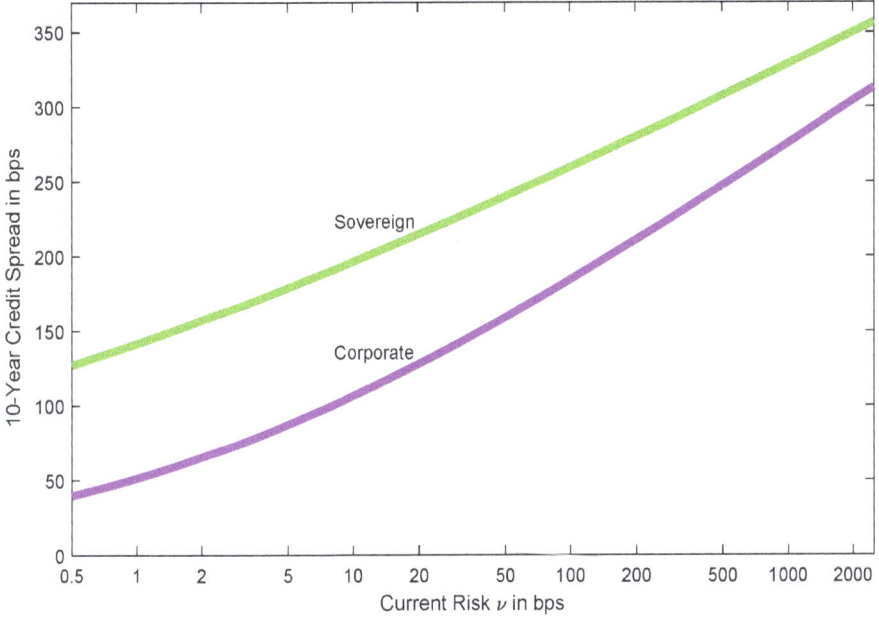

Figure 13.3: Scatter plots for 10-year credit spreads and current $\nu = E$

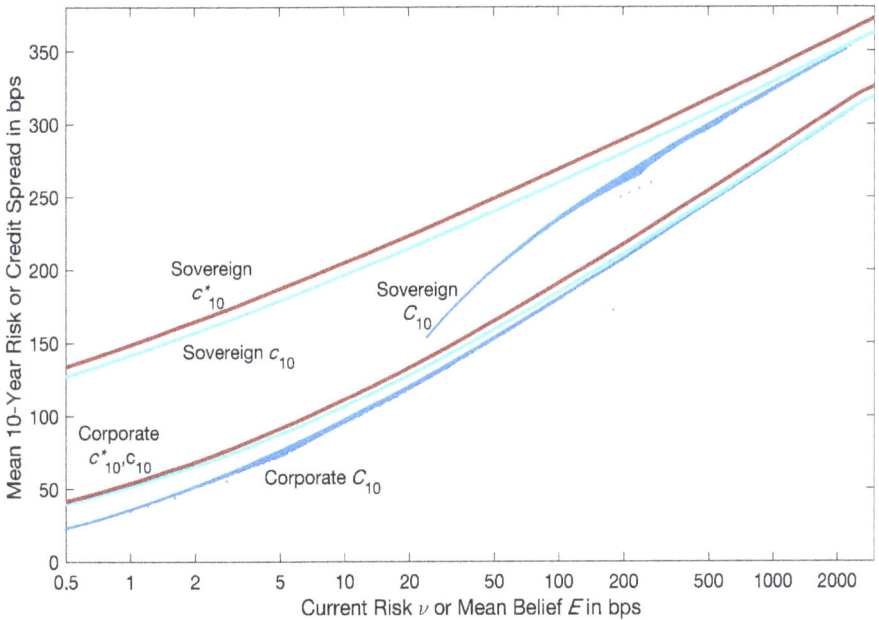

sovereign credit history, not just periods of high risk, their analysis of lender attitudes focused on periods just prior to default. This self-selects for trouble. In benchmark simulations, the average sovereign $v = 712$ bps on the eve of default is nearly 60% higher than the corresponding average $E = 450$ bps.

More generally, Reinhart and Rogoff contended that sovereign debt carries an unjustified mystique, with lenders far more prone to understate sovereign risks than private borrower risks. However, big differentials occur with Perfect Learners too, who don't worry about mystique. On the eve of corporate default in our benchmark simulations, the average $v = 424$ bps is less than 10% higher than the average $E = 391$ bps. Why is the corporate risk gauged so much better? The answer is simple. The corporate pool is much larger, which permits far more relevant sampling.

Let us turn now to measures of complacency and surprise. Figures 13.4(a)-(b) plot 100-year Monte Carlo simulations of 10-year spreads for corporate and sovereign benchmarks. Each chart compares C_{10} given beliefs about v to the ideal spread c_{10} if v were known. While no 100-year stretch can capture the full variety of responses, I have picked samples with mean and standard deviation close to simulation averages.

The tracking in Figure 13.4(a) for the corporate benchmark looks excellent, as deviations seem relatively small and rarely persist more than a few months. The tracking in Figure 13.4(b) for the sovereign benchmark looks notably worse. Sovereign credit spreads often lag their ideal values for several years and fail to capture extremes.

These impressions are confirmed in 100,000-year simulations. For the corporate benchmark, C_{10} ranges 94% as widely as the ideal c_{10}. The RMSE between them is 19 bps, with half of absolute deviations under 12 bps and 1% over 50 bps. For the sovereign benchmark, C_{10} ranges only 58% as widely as the ideal c_{10}. The RMSE between them is 36 bps, with half of absolute deviations under 22 bps and 1% over 100 bps.

In practice, it is hard to compare C_{10} to the ideal c_{10} because we don't know v and its laws of motion. It is tempting to gauge rationality by measures like the jump ΔC_{10} in credit spread after a single default. Figure 13.5 displays scatter plots of ΔC_{10} versus C_{10} for the two benchmarks. Both

Figure 13.4: Simulated daily tracking for 100 years of 10-year credit spreads

(a) Corporate benchmark

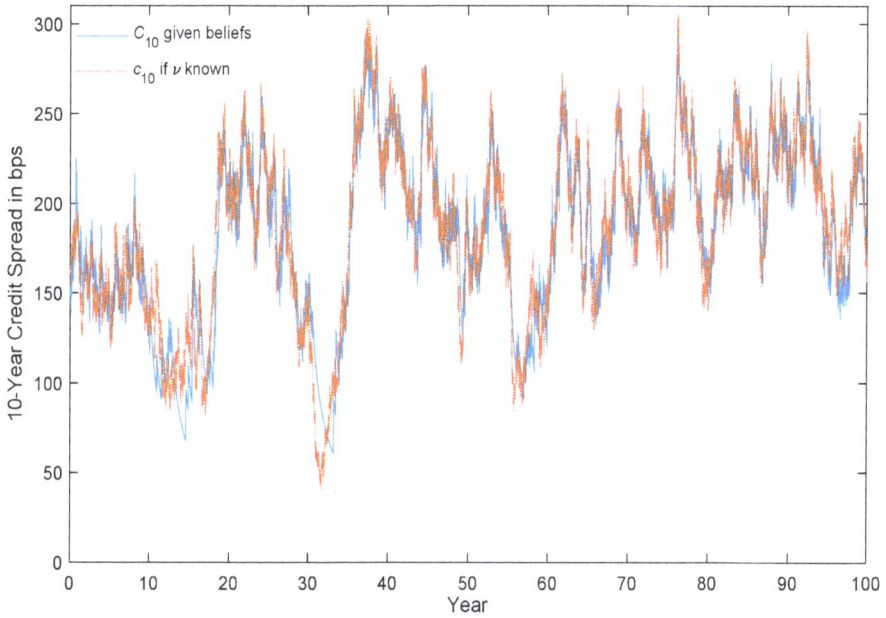

(b) Sovereign benchmark

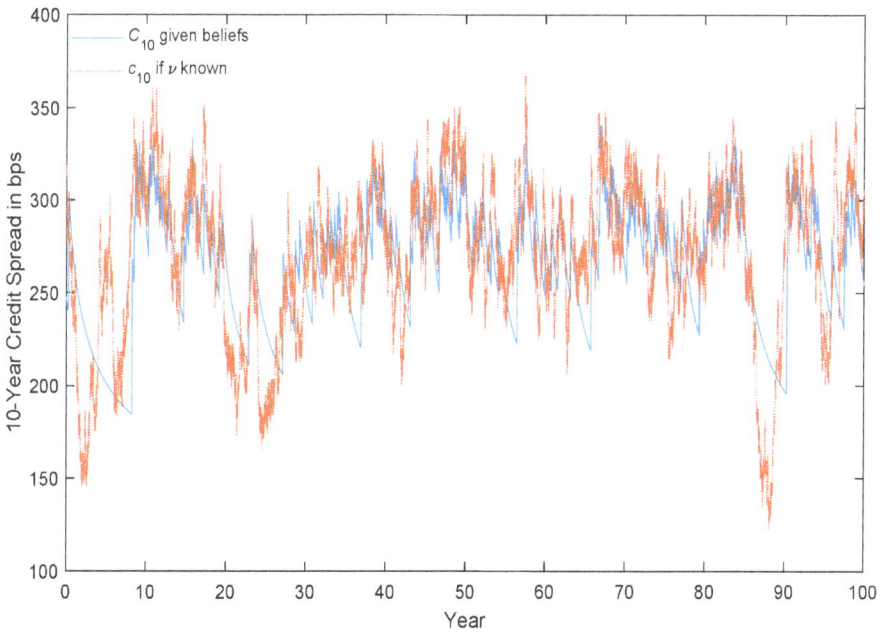

are thin, downward-sloping crescents. At low baseline C_{10}, the jumps are so high that affected markets are bound to look as if they were lulled into complacence and then shocked into overreaction.

For the corporate benchmark, 5% of defaults trigger a jump ΔC_{10} of over 20 bps and 0.7% of defaults trigger a jump of over 30 bps. The sovereign benchmark is far more sensitive, as 42% of defaults trigger a jump of over 20 bps and 4% trigger a jump of over 50 bps. Every sovereign jump lands C_{10} over the long-term mean of 264 bps, whereas no corporate jump will land a C_{10} of less than 137 bps (which occurs 24% of the time) above the long-term mean of 174 bps. A Perfect Learner could easily appear to be grossly naïve about the risks of sovereign debt.

Compared with jumps in E, most jumps in C_{10} are small. The average jump is 4.4 bps for the corporate benchmark and 21 bps for the sovereign benchmark. Bear in mind, however, that the price of a 10-year bullet declines in percentage terms by roughly 10 times ΔC_{10}. Consider a lender who holds one 10-year bullet from every issuer in the benchmark. For 2500

Figure 13.5: Scatter plots for corporate and sovereign benchmarks of jumps ΔC_{10} after a single default versus C_{10} just before default.

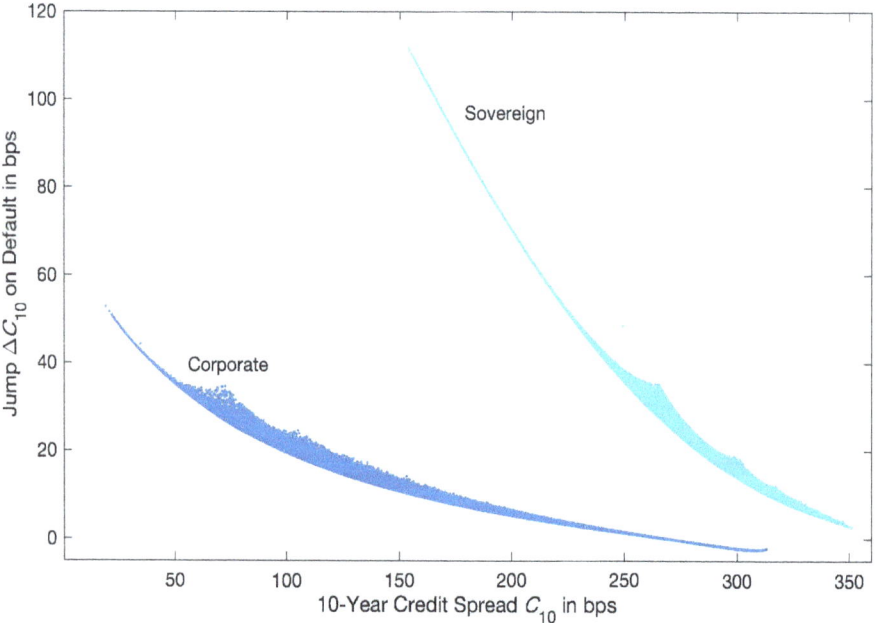

corporate issuers, a single default trims portfolio value by 4 bps directly and by an average 44 bps indirectly. For the sovereign benchmark with 80 independent issuers, a single default trims portfolio value by 125 bps directly and an average 210 bps indirectly. In both cases the indirect effect dominates the direct effect. This has important consequences for banking regulation as we will see in Chapter 15.

Markets for Bonds

In practice, the bond market will consist of many different gamblers, each with capital K_i and a view B_i of fair price. If the market price is B for a bullet paying 1 and each gambler bets to maximize long-term wealth, the Kelly criterion will induce long or short bullet holdings H_i of

$$H_i = \frac{B_i - B}{B(1-B)} K_i. \tag{13.1}$$

Provided net bond issuance is small, the market price will approximate the capital-weighted average, and capital will flow over time toward the best gamblers. Eventually the market will price bullets close to how a Perfect Learner would, even it never grasps the underlying laws of motion.

The core challenge is time. Each bullet takes 10 years to mature without default, so if updates never overlap the market will need millennia to sort things out. Fortunately, the market trades bonds of various maturities and continually reprices them. This allows lenders to infer consensus risks at all maturities and receive quicker payouts on their bets. The flip side is that the payout absent default is a shorter-duration bullet, whose price depends on others' beliefs about the future.

To illustrate this, I will replace Perfect Learners with Good Learners, who understand the drivers of v but don't know the true intensities. Let their corporate models choose between shape $E_0 = \{120,160,200,240\}$ bps, $D_0 = \{0.4,0.6,0.8\}$, $s = \{0.9,1.2,1.5\}$ and $\Lambda = \{0,0.03\}$. Each of the 72 possible combinations implies a different relation between C_{10} and the current v. As Figure 13.6 shows, the disagreements span a broad range.

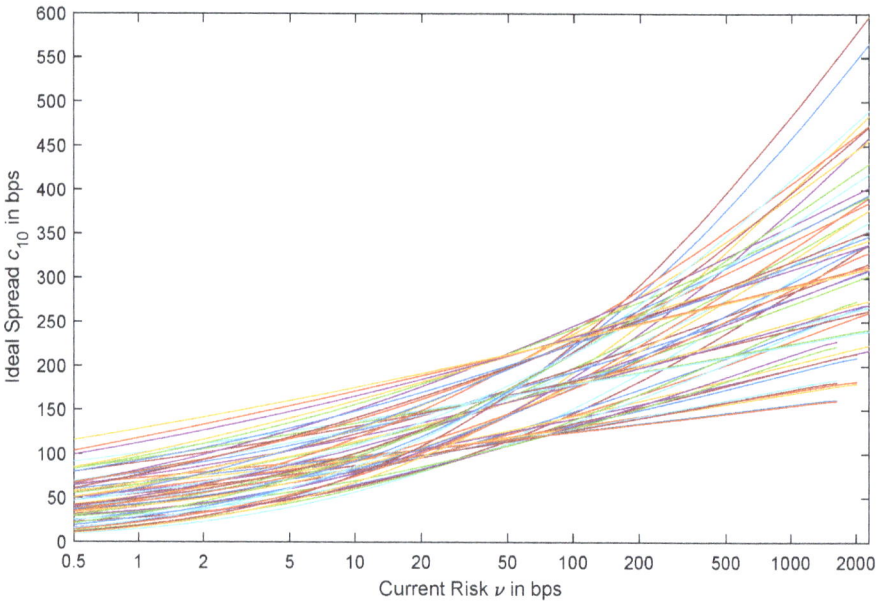

Figure 13.6: Ideal 10-year credit spreads for combinations of $\Lambda = \{0, 0.03\}$, $E_0 = \{120, 160, 200, 240\}$ bps, $D_0 = \{0.4, 0.6, 0.8\}$, and $s = \{0.9, 1.2, 1.5\}$

Since the various Good Learners will also differ in their perceptions of ν, their disagreements about fair C_{10} likely range even wider than Figure 13.6 suggests. Imagine that we embed them in an inter-threaded market, trading a market security that initially consists of one 10-year bullet for each of the N issuers. Denoting the day's default rate by δ, the next day the price of the security equals the number $(1-\delta)N$ of non-defaulting issuers times the revised market price of a 10-year-minus-one-day bullet. Imagine that every security holder proceeds to close out all one-day-old bullet positions and to reinvest wealth in new 10-year bullets, after applying (13.1) with new capital shares and new views on fair price.

In simulations of the corporate benchmark, the Good Learner ITM tracks the Perfect Learner very well in most respects. With doubts of up to 0.1 per year, the correlation in C_{10} appraisals usually exceeds 0.995, as does the correlation in daily ΔC_{10}. However, like the bots in Chapter 11, the Good Learner market has trouble identifying the long-term mean. While it steers away from extremes, its judgment is coarse and lets moderate biases persist longer term.

Suppose we start with equal weights on $E_0 = \{160,200,240\}$ bps, so that the market initially overstates the long-term corporate mean by 20 bps. Even after 1000 years, the ITM C_{10} tends to overshoot the Perfect Learner C_{10} by roughly 10 bps. Alternatively, if we start with equal weights on $E_0 = \{120,160,200\}$, the bias reverses with a similar undershoot. Daily trading with 10-year bullets rarely mitigates bias much better than strategies that buy and hold until maturity.

While bond trading at a host of different maturities seems likely to speed learning, long-term fair spreads cannot be identified nearly as fast as short-term fair spreads can. As we saw in Chapter 11, the core problem is the time needed to collect independent samples of long-term means. Overlapping observations are too correlated to provide much extra insight, and taking them more frequently just compounds the correlation. This doesn't mean that bond prices aren't useful. They're great measures of consensus beliefs. They're great at incorporating new information. They're just not great at predicting long-term divergence from short-term trends, due to huge uncertainty about what might happen along the way.

Another caveat concerns risk premia. Unlike at Harold's Casino, where net holdings are zero, net bond holdings are positive. Lenders must expect a mean positive return to be willing to hold them. If ϕ denotes the percentage discount for risk and h denotes the net bond stock as fraction of total capital K, then

$$(1-\phi)B = \sum_i \kappa_i B_i \quad \text{and} \quad (1-\phi)\sum_i H_i = hK.$$

If each gambler applies the full Kelly strategy (13.1), equilibrium requires

$$\phi = h(1-B).$$

Since ϕ is positive and fluctuates with v, spreads will be both higher and more volatile than our previous calculations suggest. The impact is guaranteed to be small only if h is tiny or if long-term default risk is negligible.

14

Credit Grades

To what extent is it appropriate to treat credits as iid? Formally, iid requires equal risk and independent outcomes. Informally, it allows some wobble. Still, application to corporate and sovereign debt stretches the notion to extremes. It is better to pool only credit risks that have similar orders of magnitude. These are known as credit grades and specialized firms have arisen to rate them. By convention, most credit rating agencies assign at least seven distinct letter grades, with each step down in grade at least tripling the estimated default risk. They also subdivide major letter grades into three tiers, which are identified by numbers or signs.

Table 14.1 displays average one-year global corporate default rates by credit grade for the two most comprehensive series I found. The data from Moody's (Ou et al 2018, Exhibits 30 and 32) cover 1920 through 2017. The data from S&P (Vazza and Kraemer 2018, Tables 3 and 4) cover 1981 through 2017. As the data can be weighted either by year or by number of issuers, I applied both weightings and averaged their results.

Both lists exhibit roughly geometric spacing with several thousand-fold differential across the whole. Despite the different time periods, they roughly coincide except for the lowest grades, where S&P appears to set higher thresholds. For our analysis, I shall merge the pairs into grades renamed triple-A, double-A and so on.

Table 14.1: Average annual global corporate default rates
by letter credit grade in basis points

Moody's Rating	Default Rate 1920–2017	S&P Rating	Default Rate 1981–2017
Aaa	0	AAA	0
Aa	6	AA	2
A	9	A	6
Baa	26	BBB	19
Ba	112	BB	89
B	329	B	397
Caa-C	1030	CCC-C	2544

Figure 14.1 charts average default rates by tier on a log scale. The rates give equal weight to the longest samples available, which came from S&P for 1981-2017 and Moody's for 1983-2017.[10] The best log-linear fit accounts for 98.5% of the variance. Its slope of 0.48 per tier implies an average risk multiplier of $e^{3\times0.48} \approx 4.2$ per drop in letter grade. The confidence interval spans 0.44 to 0.51 for a risk multiplier of 3.8 to 4.6.

Since the aggregate risk v fluctuates considerably over time, each credit tier seems best viewed as an average multiplier ω applied to v. With $\langle v \rangle \approx 180$ bps, the estimated Baa3/Ba1 border of 44 bps implies an investment/speculative threshold for ω of 0.24. Similar computations estimate a maximum ω of 0.003 for triple-A, 0.014 for double-A, 0.057 for single-A, 1.02 for double-B, 4.29 for single-B and 18.1 for triple-C.

Despite the steady gradient, capital markets place special emphasis on grades of triple-B or higher. These are known as investment grades, which appeal to a broader class of lenders than the speculative grades below. Firms prized investment grade ratings and tended to withdraw from ratings (which they typically pay for) if they fell too far away. In recent decades, opportunities for higher risk financing have greatly expanded, causing many higher-risk firms to seek ratings. Nowadays roughly half of ratings are speculative. Still, investment grades carry extra cachet and firms near the boundary strive hard to obtain them. Since 2000, both triple-B and single-B ratings have significantly outnumbered double-B.

Figure 14.1: Average default rates per tier for Moody's/S&P mix 1981-2017

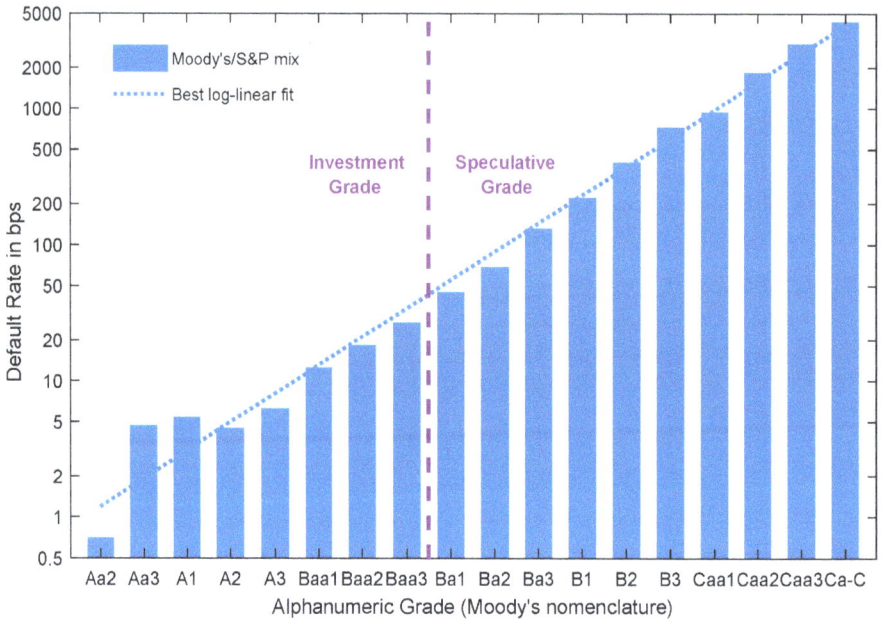

The main practical justification for special treatment of investment grades is that they hardly ever default at high rates. In Moody's data spanning nearly a century, its lowest investment grade has never experienced a year with 200+ bps default rate. In contrast, its highest speculative grade has experienced four years with 400+ bps default rates and one year with a default rate of 1155 bps. Since 1981, the maximum corporate default rate for the lowest investment tier was 194 bps for Moody's and 133 bps for S&P, compared to a maximum default rate for the highest speculative tier of 505 bps for Moody's and 370 bps for S&P.

Long-term investment-grade loans are riskier than these numbers suggest, since debtors' credit rating might decline before their debt matures. A common regulatory response distinguishes between lower investment grades where risk of deterioration to speculative grade seems significant and higher investment grades where it does not. Banking regulation has often required negligible capital buffers (reserves against unexpected losses) for double-A or triple-A loans, moderate buffers for single-A loans, and significantly higher buffers for speculative-grade loans.

As the next chapter explains, there are serious drawbacks to this approach. Nevertheless, its influence increases the importance of estimating ω well and forecasting its likely evolution.

How do we know which value of ω to assign to a given credit at a given time? Ideally, we find sound indicators z of relative default risk, pool all credits with similar values of z, count the relevant number D_{eff} of defaults in T_{eff} years of relevant observations, and compare the default rate D_{eff}/T_{eff} to the aggregate $\langle v \rangle$. The best fitting ω will equal their ratio, with a relative standard deviation of $\sqrt{1/D_{eff}}$.

Good ratings should nearly always assign the credit to the correct tier or a tier next to it. For tiers not quite 0.5 log units wide, the relative standard deviation should be less than 25%, which requires $D_{eff} > 16$ and $T_{eff} > 16/(\omega\langle v \rangle)$. Assuming the relation between ω and z is broadly stable for three decades, this requires on average at least $1/(2\omega\langle v \rangle)$ observations per tier per year.

Do we have enough corporate data to achieve that? Near the investment/speculative border, we need on the order of 100 observations per year, which each side easily provides. As grades deteriorate, ω mounts so fast that again the answer is yes. Higher grades are more challenging. Good identification of the middle single-A tier requires roughly 1000 observations per year, which approaches the size of the entire single-A grade. For the lowest double-A tier we need at least 2000 observations per year, when the entire grade offers only have a few hundred. From this perspective, low ω seems doomed to be highly uncertain. The left side of Figure 14.1 hints of this, as the default rates for Aa3 through A3 are close despite a four-fold difference in estimated risks.

If we can find a statistical relationship between z and ω that spans a broad range of values, estimating low ω is much easier. For a best-case example, imagine credit analysts know that $\omega = az^b$ for a precisely measured scalar z, that they draw on 30 years of data from the corporate benchmark for v and that the distribution of ω is broadly consistent with the distribution of Moody's corporate ratings in 2000 (Ou et al 2018, Exhibit 41).[11] Figure 14.2 charts the relative RMSE in rated ω versus actual ω for 50 different simulations, with the horizontal axis labeled by credit grade.

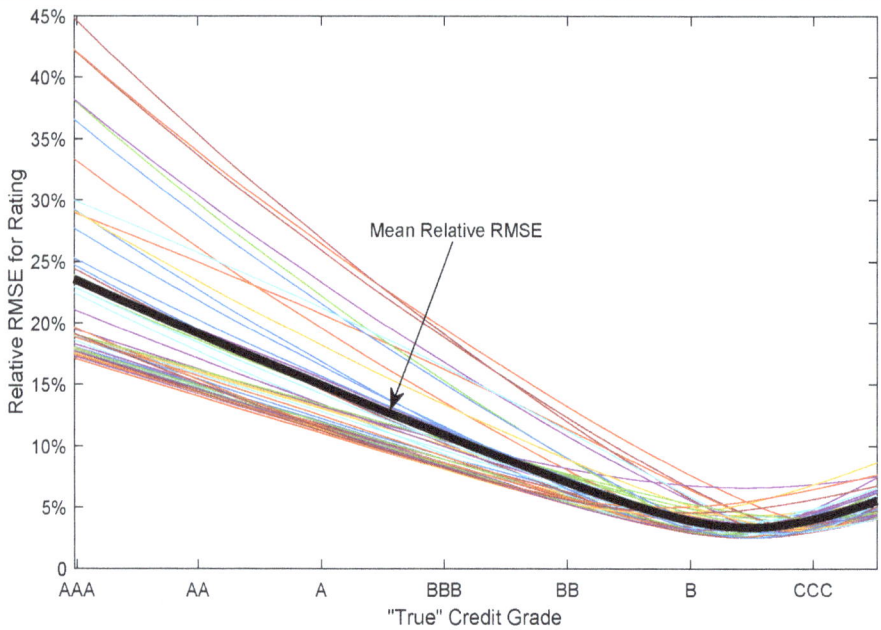

Figure 14.2: Relative RMSE in rated ω for 50 simulations with 30 years of evidence from 2500 corporates, $\langle v \rangle = 180$ bps and ω per Moody's in 2000

Low risks are consistently rated much less reliably than the rest. Still, enough information percolates from riskier credits to identify nearly all tiers. The only exception is the triple-A/double-A divide.

The model treats default like the radioactive decay of an identifiable "riskium" element $r \equiv z^b$. If riskium really existed, we wouldn't need to examine any A-grade servicing to appreciate their risks, since B-grade and C-grade risks are much easier to estimate and would tell us everything we need to know. Also, a tight riskium relation implies very high correlations of annual default rates across most letter grades. In reality the empirical correlations never reach 0.75 and decline steeply with distance between grades.[12] Ratings practice implicitly acknowledges the limitations, as the highest triple-A grade is not tiered and its distinctions from double-A are rarely deemed significant unless they impact regulation.

Identification gets even more challenging when we turn to sovereign credits. To put a lower bound on our uncertainty, suppose 80 sovereigns are observed for 60 relevant years each, with average $\langle v \rangle = 270$ bps per our

Figure 14.3: Relative RMSE in rated ω for 50 simulations with 60 years of evidence from 80 sovereigns, $\langle v \rangle = 270$ bps and ω per Moody's in 2000

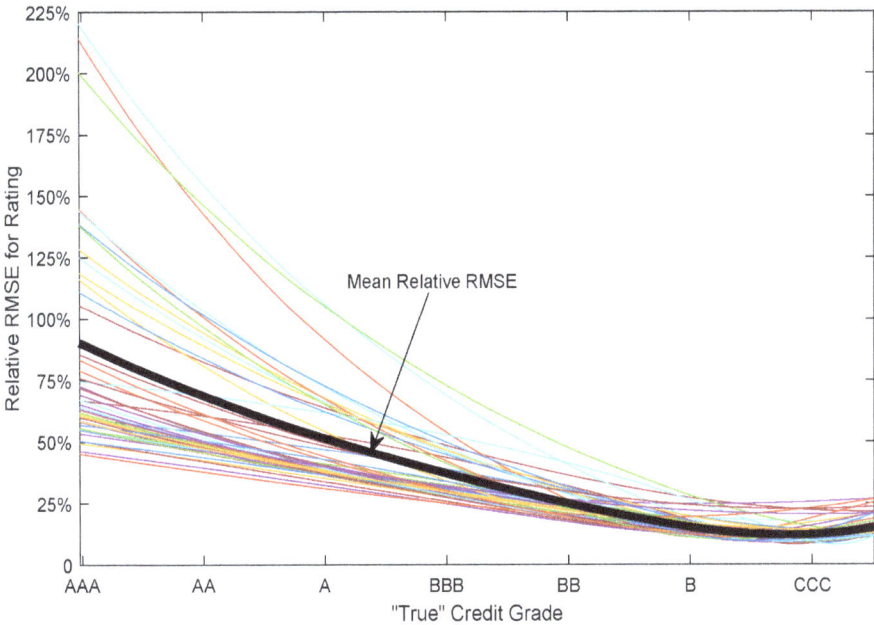

benchmark and with ω distributed like Moody's sovereign credit ratings in 2000 (Liu et al 2018, Exhibit 8).[13] Figure 14.3 charts the relative RMSE for 50 different simulations. The average relative RMSE is on the order of 80% for triple-A, 65% for double-A, 50% for single-A, 35% for triple-B, and 25% for double-B. All the investment grade RMSEs exceed the 25% upper bound for confident assignment of tiers. The worst-case errors are much wider.

The extra wobble reflects the small size of sovereign pools. A single default for a sovereign A-grade z ups the estimated risk associated with it by at least 20% while slightly lowering the estimated risk for C-grade z. Judging from Figure 14.3, investment grade credits will often be assigned to the wrong tier and doubts in the highest ratings might span a full grade. Again, this presumes a perfect riskium relation, where a Greece that is 50 times as likely to default as Germany provides 50 times more information about German default risks than Germany does. When we acknowledge that such relationships are muddied by huge differences in context, even the differentiation between sovereign A-grades seems suspect.

Credit Migration

Let us next analyze migration across credit ratings. Table 14.2 displays Moody's average annual transition rates by letter grade for 1920 to 2017 (Ou et al 2018, Exhibit 25). Defaults are categorized separately as are credits that withdraw from ratings. We can view the table as a transition matrix P, whose elements p_{ij} approximate the probability that a credit starting in category i ends the year in category j.

For modeling we need a smaller matrix Π that merges the last two columns into the rest. Its elements π_{ij} estimate the probability that a risk multiplier shifts from near-ω_i to near-ω_j, whether or not the credit exits ratings through withdrawal or default. How should we convert P into Π? One method assigns exits wholly to grade at the start of the year. Another assigns them in proportion to grade at the end of the year absent exit. The

Table 14.2: Average 1-year transition rates in percent for corporate credits 1920-2017 from Moody's (DF=default, WD=withdrawn)

To: From	Aaa	Aa	A	Baa	BB	B	Caa	Ca-C	DF	WD
Aaa	87.71	7.94	0.58	0.07	0.02	0.00	0.00	0.00	0.00	3.67
Aa	0.82	85.15	8.51	0.42	0.06	0.04	0.02	0.00	0.02	4.95
A	0.05	2.46	86.78	5.37	0.48	0.11	0.04	0.01	0.05	4.64
Baa	0.03	0.14	4.12	85.72	3.79	0.69	0.15	0.02	0.17	5.17
Ba	0.01	0.04	0.42	6.12	76.32	7.17	0.71	0.11	0.88	8.22
B	0.01	0.03	0.14	0.45	4.78	73.49	6.62	0.52	3.27	10.70
Caa	0.00	0.01	0.02	0.08	0.34	6.51	67.87	2.85	7.96	14.35
Ca-C	0.00	0.00	0.05	0.00	0.56	2.29	8.94	39.39	26.66	22.12

method I favor comes close to their average. For credits that start the year in grade i, their mean time t_{ii} in i during the year is close to $\frac{1}{2}(1+p_{ii})$ while their mean time t_{ij} in any other grade j is close to $\frac{1}{2}p_{ij}$. Let u_i denote the yearly exit rate for a credit currently in grade i After solving the equations $\sum_j t_{ij}u_j = p_{i,default} + p_{i,withdraw}$ for the various u_i [14] we compute

$$
\begin{aligned}
\pi_{ii} &= p_{ii} + t_{ii}u_i = p_{ii} + \frac{1}{2}(1+p_{ii})u_i \\
\pi_{ij} &= p_{ij} + t_{ij}u_j = p_{ij} + \frac{1}{2}p_{ij}u_j
\end{aligned}
\tag{14.1}
$$

Table 14.3 displays the conversion results for Table 14.2. Applying (14.1) to the main alternative data sets generates very similar estimates. Using Moody's transitions by grade for 1970-2017 or averaged transitions by tier for 1983-2017 (Ou et al 2018, Exhibits 26 and 29), the estimated Π rarely differ by more than 100 bps. Using S&P transition rates by grade for 1981-2017 (Vazza and Kraemer 2018, Table 21), the only major change in Π is faster reversion at the low end: a 14% conditional migration rate from the merged C grades to single-B.[15]

By taking powers of Π, we find an average long-term $\langle \omega \rangle \approx 5$ absent default and withdrawal. Default and withdrawal lower $\langle \omega \rangle$ to about 2, which means that the average risk per rated corporate credit is about twice the average rated corporate risk per year. Why? Because high ratings persist much longer than low ratings: e.g., averages of 18 years for A-grade versus 3 years for C-grade.[16] Figure 14.4 gives the tiered breakdown for 1983-2017. The estimates by grade for 1920-2017 are very similar.

Sovereign withdrawals are much rarer than corporate withdrawals, so their raw transitions P look quite different from corporate P and their rating lives are much longer. However, the sovereign Π inferred from S&P 1975-2017 data (Witte and Ontko 2018, Table 4) or Moody's 1983-2017 data (Liu et al 2018, Exhibit 15) is remarkably close to the corporate Π. Perhaps this is coincidence. In recent decades a host of less developed countries, like weaker corporations, have sought ratings in hopes of improving their access to global markets. It is hard to gauge the long-term trends as there are only about 130 rated sovereigns and less than 50 have been rated for more than 25 years.[17]

Table 14.3: Estimated yearly migration rates Π across credit grades for Table 14.2, applying (14.1)

To:	Aaa	Aa	A	Baa	BB	B	Caa	Ca-C
From:	As Percent of Rated Corporates Not Defaulting or Withdrawn							
Aaa	90.93	8.02	0.81	0.19	0.03	0.00	0.00	0.00
Aa	1.07	89.99	7.95	0.75	0.17	0.05	0.01	0.01
A	0.07	2.78	90.52	5.77	0.68	0.13	0.04	0.01
Baa	0.04	0.24	4.32	89.62	4.83	0.78	0.14	0.02
Ba	0.01	0.07	0.50	6.38	84.78	7.40	0.75	0.11
B	0.01	0.04	0.16	0.64	5.91	85.71	6.95	0.58
Caa	0.00	0.01	0.03	0.12	0.54	7.26	88.31	3.73
Ca-C	0.00	0.02	0.10	0.04	0.61	3.05	9.20	86.99

Figure 14.4: Average corporate rating life by tier for Moody's transitions 1983-2017

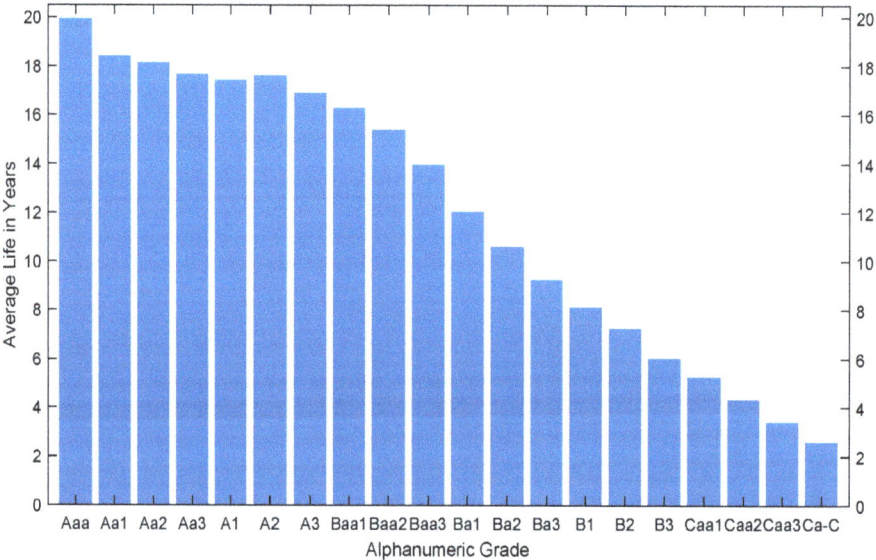

15

Value at Risk

Banks arose to facilitate payments. Since total value of bank accounts far exceeds current payments, banks can profit from lending some of the excess. Since borrowers want minimal pressure to repay early while depositors want minimal constraints on withdrawals, banks potentially profit even more from letting the average term of loan assets stretch well beyond the average term of bank liabilities. This is known formally as duration mismatch and informally as borrowing short to lend long.

Duration mismatch induces credit expansion, which bolsters the economic appeal. However, it risks destabilizing the payment system through bank losses on long-term loans. Even fear of such losses can trigger a chain reaction of deposit withdrawals and bank closures.

To mitigate these risks, banks are required to hold capital buffers. These were traditionally set through caps on total leverage on the grounds that any loan might prove worthless. However, risks vary greatly by credit grade. An alternative approach aims for enough capital buffer to cover all losses from default nearly all of the time. A favored threshold is 99.9% per year, or one breach expected per thousand years. The capital needed to provide this protection is known as the Value at Risk or VaR.

Since VaR covers only the least loss among the worst cases, a more accurate name would be Value Not at Risk. What's at risk is everything else. Unfortunately, the worst aspect of VaR isn't the dubious semantics. It's that major breaches occur far more often than standard models predict. This chapter will explain why.

Let's start with the simplest VaR estimator. Consider a portfolio of N equally weighted iid bonds, each with risk multiplier ω on an aggregate default risk v. For large N the 99.9% worst loss is approximately $3\sqrt{\omega v / N}$ below the mean. Since each drop in letter grade is associated with a quadrupling of ω, it appears to double VaR.

We can see traces of this approach in the 2004 Basel Accords for international banking, better known as Basel II. Table 15.1 lists the standardized capital buffers recommended by Basel II. [18] The buffers approximately double for each drop in letter grade in the middle of the range. However, no VaR-based theory can fully justify Table 15.1. The values seem to be pragmatic compromises between VaR-type calculations and prior norms.

Let's now estimate VaR for our benchmark models. The fluctuations in risk cause defaults to cluster, which induces higher VaR for the same average ωv. Suppose a bank holds a portfolio of benchmark debt with equal weighting across all N members of the pool, whose risks are assumed to be conditionally iid. The debt instruments are T-year bullets, which are swapped the next day at fair market price for new T-year bullets in order to keep duration constant. When default occurs, a new issuer with equal risk enters the market and the portfolio is rebalanced to maintain a $1/N$ share with each issuer.

Table 15.1: Basel II standardized capital buffers

	Double/ Triple A	Single A	Triple B	Double B	Single B	Any C	Un-rated
Sovereign	0	1.6%	4%	8%	8%	12%	8%
Bank	1.6%	4%	8%	8%	8%	12%	8%
Corporate	1.6%	4%	8%	8%	12%	12%	8%

Given a stable credit spread, every non-defaulting bond appreciates overnight as there is one less day of default risk. When no bonds default, the credit spread declines unless it is already at a minimum. When default occurs, defaulted bonds are presumed worthless and the credit spread rises on the rest. My model prices all these influences along with anticipated diffusion or reset. For simplicity I ignore interest charged for time delay or risk aversion, whose impact on VaR is small.

The drawdown over the next year is the highest percent loss from the starting point. Simulations for 50 trading weeks per year for a million and one years allowed the computation of 50 million yearly drawdowns. I sorted the drawdowns by their starting E and then subdivided them into 250 non-overlapping sets of 200,000 points each. In each set, I estimated the 99.9%-ile, 99%-ile and 95%-ile VaR for the mean starting E as the 200th worst drawdown, $2,000^{th}$ worst drawdown, and $10,000^{th}$ worst drawdown respectively Figure 15.1 displays the three sets of VaR for 10-year bullets drawn from the corporate benchmark.

Figure 15.1: VaR for portfolios of 10-year bullets from corporate benchmark

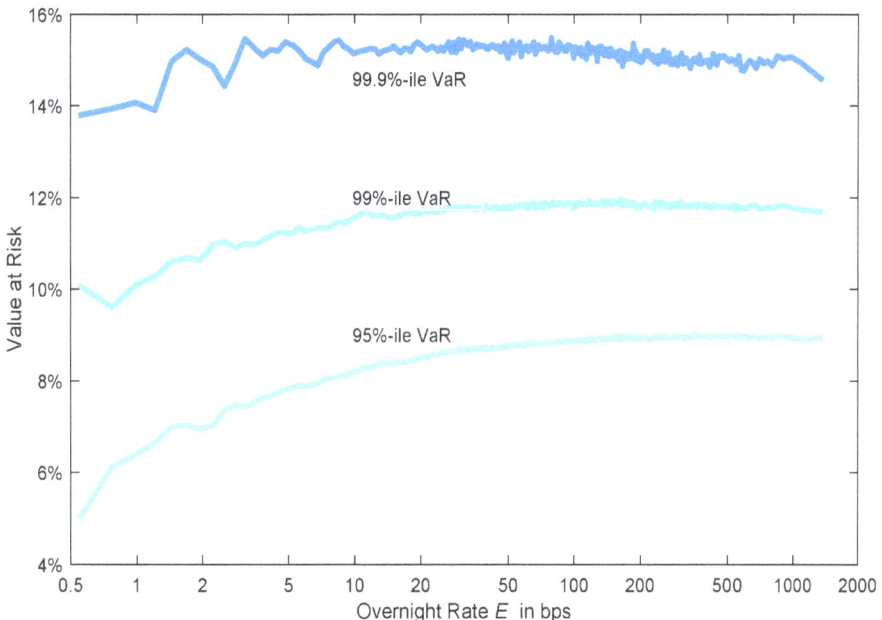

Even a million-year simulation cannot flatten all the random fluctuations. Nevertheless, the general trends are clear. VaR thresholds don't vary much with current risk E except in the highest investment grades. The most striking feature is the high average: 15.1% for the 99.9%-ile, 11.7% for the 99%-ile, and 8.7% for the 95%-ile. At first glance, only the 95%-ile VaR looks broadly consistent with the buffers in Table 14.1 and only for grades below single-A.

To help reconcile the two approaches, let's acknowledge that defaulting credits typically repay some of the principal with lags. Incorporating a $Z\%$ net salvage value reduces VaR for a bullet by $100 - Z\%$. At a salvage value of 40%, which is in the ballpark of reported historical norms, the 99.9%-ile VaR comes much closer to Basel II standards. However, since salvage values are highly variable, VaR should factor in the likelihood that post-default payouts are low. VaR calculations should also recognize that average credit $\langle \omega \rangle$ exceeds 1 and that benchmark portfolios for corporates and sovereigns behave differently. Figure 15.2 displays the 99.9%-ile VaR for various alternatives.

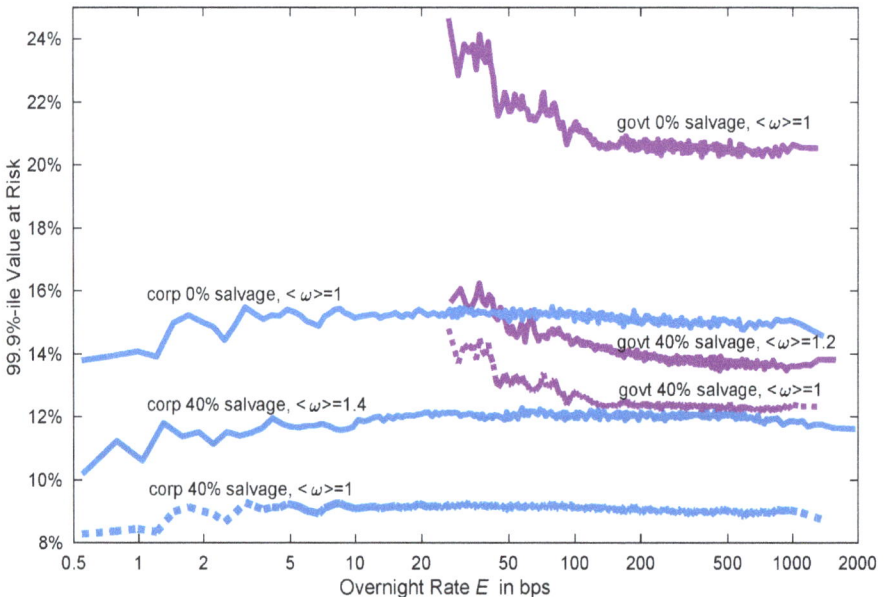

Figure 15.2: 99.9%-ile VaR for portfolios of 10-year bullets from corporate and sovereign benchmarks

VaR thresholds get higher as bond duration rises and lower as bond duration declines. The differentials are greatest when E is lowest because it takes time to migrate from low risk to high risk. Given a long enough wait, low E brooks greater downside than high E since the bonds have farther to fall. When E is high, the differentials are small, as the main VaR risk is an extreme concentration of defaults and that can occur at any duration.

Figure 15.3 charts the 99.1%-ile VaR for various durations of bullet portfolios from the corporate benchmark. Its most striking feature, which is robust to most adjustments to salvage values or average risks, is the huge dependence on duration. Basel II missed this. While it demanded ample capital to cover the direct losses from an unexpected cluster of defaults, it ignored the consequent pessimism about v and depreciation of longer-term bonds.

For deeper exploration, we need to introduce multiple credit grades and migrations across grades into our simulations. Here is how I estimated fair values for bullets of each starting grade:

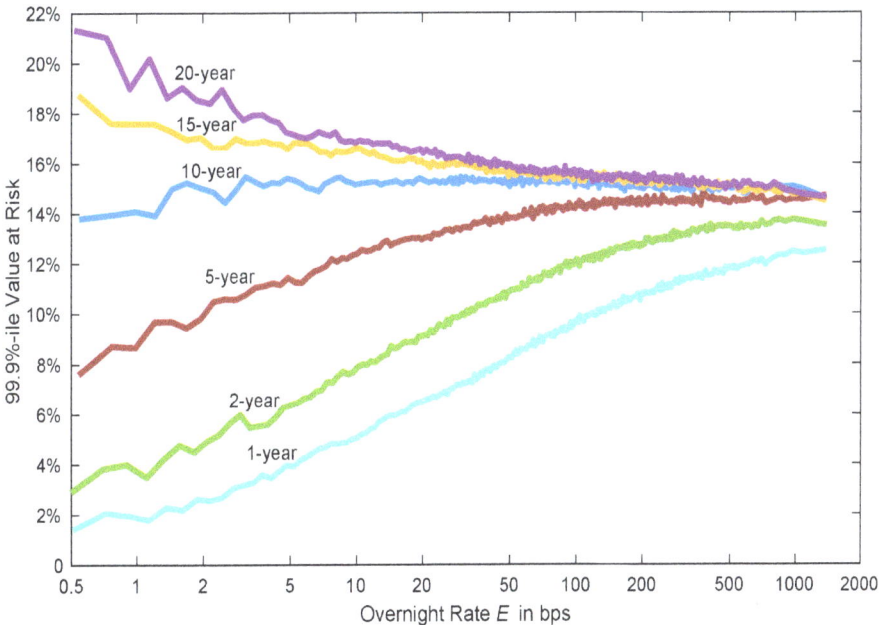

Figure 15.3: Impact of duration on 99.9%-ile VaR for benchmark corporate bullets, assuming zero salvage value and $\langle \omega \rangle = 1$

- Assign each grade a risk multiplier ω equal to the geometric mean of its range, after fitting a straight line to Figure 14.1.[19]
- Convert the yearly transition rates in Table 14.3 to weekly rates.[20]
- Compute survival probabilities for 10-year bullets, assuming ω and ν evolve independently.

Figure 15.4 plots the survival probabilities S_{10} as functions of current ν. For A-grade credits, S_{10} exceeds 0.97. For triple-B credits, S_{10} exceeds 0.92. Survival rates for speculative-grade credits are much lower, especially when the initial ν Is high.

It is evident from Figure 15.4 that the biggest drawdown risk for A-rated credits involves a slip to speculative grades, whereas the biggest drawdown risk for speculative grades involves a mixture of downgrade and surge in ν. However, Figure 15.4 can't tell us the intensity of rare surprises. To estimate VaR, we need to track fair values of portfolios of M bonds that start in the same credit grade G but might default or migrate to different grades. Here is the procedure I followed:

Figure 15.4: 10-year survival probabilities for benchmark corporate credits

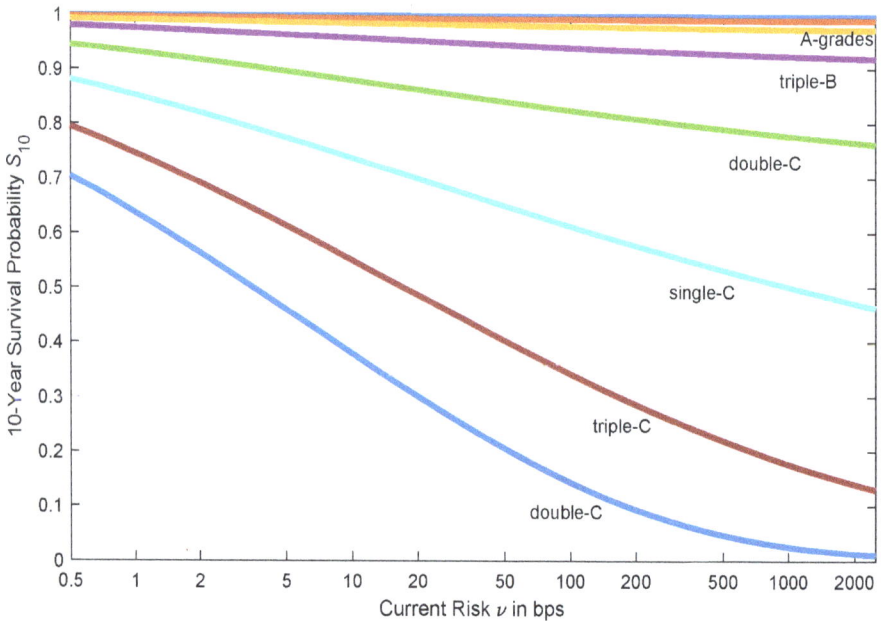

- Simulate a million years of weekly evolution of v and aggregate defaults for the corporate benchmark.
- Assume the market updates beliefs about v like a Perfect Learner by inference from aggregate defaults.
- Form at the start of every year a portfolio of M 10-year bullets, with each bullet bearing the same initial risk multiplier ω_G but issued by a different borrower.
- Hold onto the portfolio for exactly one year, during which time each credit is prone to random independent change in ω with transition probabilities Π and random independent defaults with probability ωv.
- Price each portfolio fairly every day assuming current ω are always identified correctly, that Π is known, and that defaulted bonds have no salvage value.
- Calculate each year's drawdown as the maximum percentage price drop during the year.
- Calculate the 99.9%-ile VaR as the 1,000th biggest drawdown out of the million drawdowns.

Portfolio diversity is often measured as the ratio of variance of returns from a single asset to the variance of the returns from the portfolio.[21] Under the assumptions above of equal initial weights, independent transitions in grade, and conditionally independent defaults, the diversity equals the pool size M. In contrast, assets that are ρ-correlated with each other can never form a portfolio with diversity much above $1/\rho$. Basel II posited correlations of 0.04 for retail exposures, 0.15 for residential mortgages, and 0.12-0.24 for corporate exposures, which implied an effective diversity between 4 and 25. Since M in my simulations corresponds more to diversity than to the nominal exposures, my investigation focused mostly on double-digit M.

Figure 15.5 displays 99.9%-ile VaR across different initial credit grades for M ranging from 5 to 100. There is some noise despite massive sample size. Holding M constant, the scaling is roughly linear in the initial ω or log-linear in credit grade. VaRs for speculative grades grossly exceed any

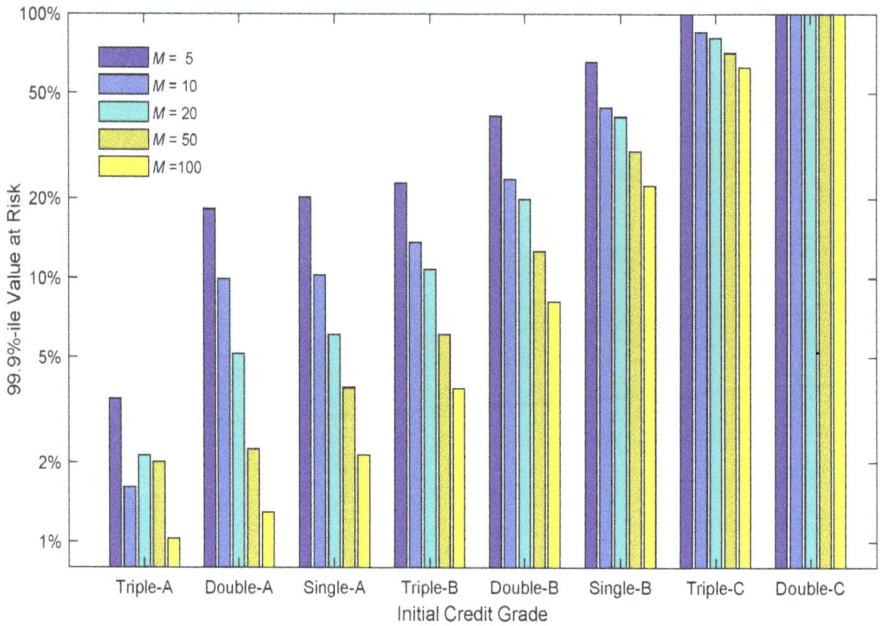

Figure 15.5: 99.9%-ile VaR for portfolios of $M = \{5,10,20,50,100\}$ conditionally independent 10-year bullets from the same initial credit grade

buffers that Basel II envisioned. Apart from triple-A, VaRs for investment grades are broadly in line with Basel II buffers only when M is a few dozen or more. Even triple-A VaR exceeds 5% when $M = 5$.

Duration-Matched Banking

Our analysis points to three major flaws in Basel II's standardized capital buffers. First, a large cluster of downgrades might impel major markdowns on previously A-grade portfolios. Second, buffers for speculative grades were too small to handle a sharp surge in the common risk v. Third, ratings of high-grade ω are bound to be highly uncertain. Were these flaws tested jointly, a Basel II-regulated banking system would almost surely plunge into crisis. Scrambling to raise capital, banks would drastically cut back their lending, pushing some borrowers into bankruptcy and depressing asset prices in the sectors they had lent to. The contraction would raise v even higher and aggravate the need for bank capital.

The financial crisis of 2008, triggered by selloffs in previously robust housing and mortgage markets, provides a case study. For over 50 years prior, US housing had avoided synchronized crisis. Most downturns were modest, which encouraged borrowers to service mortgages faithfully and gave lenders high salvage values when they did not. Exceptions involved a few occasionally troubled regions or sectors. Senior tranches of diversified mortgage portfolios sailed through historical backtests with little impairment. Basel reforms accordingly encouraged banks to take on massive senior mortgage risks and sell off the tails. One unintentional side effect was reduced pressure to monitor borrowers' ability to repay.

Easier borrowing, amplified by political pressures to expand the ranks of home buyers, boosted housing demand and prices. Rising prices reduced default risks and encouraged even more leverage, helping the appreciation feed upon itself. Like most credit-fueled booms, the news eventually got too good to stay true. As prices peaked, many mortgages were exposed as "liars' loans", backed more by expectations of future appreciation than by borrowers' ability to pay out of current incomes. As lenders recoiled, housing prices plunged and mortgage securities were downgraded in mutual reinforcement.

Several other countries experienced roughly parallel mortgage booms and busts. This exacerbated a related crisis in sovereign PIIGS debt—Portugal, Italy, Ireland, Greece and Spain. These crises have been blamed variously on fraud, political pressures and rating agency laxity. Yet such pressures could not have gained such sweeping force without the driver of Basel II-type banking regulation. Its focus on the good credit ratings of mortgages and PIIGS debt undermined the safety it ostensibly sought to promote.

Basel III rightly de-emphasized most grade-related capital buffers, added countercyclical corrections and re-imposed restrictions on total leverage. However, high-grade sovereign debt is still presumed to bear no risk, and the European Central Bank treats PIIGS debt similarly by de facto guaranteeing it. Given the differential tightening, lending to high-rated sovereigns is relatively more favored than before. This helps governments enlist banks to finance their debt. If duration mismatch were limited, sovereign borrowing would be kept on a tighter leash, which would promote more fiscal responsibility.

Is duration mismatch crucial to financing long-term investment? No. Bank deposits played little role in financing the Industrial Revolution in Britain (Allen 2009, Chapters 6-9; Mokyr 2008, Chapters 11-12); the banks that invested operated more like modern-day venture capitalists (Brunt 2006). Life insurance companies have long-term obligations that they look to repay through long-term investments, particularly in commercial real estate. Equities have plenty of takers despite their lack of nominal maturity. Financing through duration mismatch is facing competition even in areas where banks are prominent. Private equity firms are taking over the financing of promising startups. Peer-to-peer lenders process online applications for small business and consumer loans, subdivide loans into slivers, and redistribute mostly online to various lenders, each of whom holds the specified sliver until maturity.

The best defense of duration mismatch is its financing for shorter-term investment: working capital for businesses and durable goods for consumers. The duration of the unsecured part of these loans—that is, after netting out the price of whatever collateral is easily reclaimed and resold—is rarely more than a few months. It doesn't warrant a duration mismatch of several years or longer. Banks deserve special protection only to secure the integrity of the payment system and to keep depositors from panicking. Few reforms would assure that better than ending major duration mismatch.

16

Fickle Confidence

One of the main conclusions of Reinhart and Rogoff's study of sovereign debt crises, highlighted in its title *This Time is Different*, is that markets for sovereign debt are prone to manic mood swings. When things go well for an extended period, lenders tend to underestimate risks of crisis. They are too short-sighted even to anticipate their own vulnerability to self-fulfilling panic:

> The most commonly repeated and most expensive investment advice ever given in the boom just before the financial crisis stems from the perceiving that "this time is different" [because] unlike the many booms that preceded catastrophic collapses in the past [it] is built on sound fundamentals, structural reforms, technological innovation, and good policy.... Perhaps more than anything else, failure to recognize the precariousness and fickleness of confidence [...] is the key factor that gives rise to the this-time-is-different syndrome. Highly indebted governments, banks, or corporations can seem to be merrily rolling along for an extended period, when *bang!*—confidence collapses, lenders disappear and a crisis hits. (Reinhart and Rogoff 2009, xxxiv-xxxix)

This chapter shows that rational myopia can potentially explain most of these phenomena, even when perceptions don't affect the underlying default risks. Rational myopia can be useful from a policy perspective since it speeds recovery after crisis. However, our analysis supports Reinhart and Rogoff's warning (2009, xxv) that "excessive debt accumulation [...] often poses greater system risks than it seems during a boom". Indeed, it implicitly calls for extra vigilance by policymakers, to help check what lenders almost surely won't.

Let us start by revisiting the sovereign benchmark. I will fold aggregate risk and individual multipliers into a composite $v^* \equiv \omega v$ and model it as a mean-reverting diffusion to a gamma density. Shape $D_0 \approx 0.25$ and speed $s \approx 1.8$ fit the revision best.[22] To diminish the influence of weaker sovereigns, I will reduce the long-term mean to $E_0 = 200$ bps.

Figure 16.1 displays annual risks in a thousand-year simulation. The risk climate fluctuates a lot over time. In a million-year sample, a fifth of decades average $v^* < 10$ bps while nearly a tenth average $v^* > 500$ bps. A quarter of centuries average v^* under 100 bps or over 300 bps.

Might taking a long view be worthwhile to offset complacency or panic? Consider the following thought experiment. Imagine the market

Figure 16.1: Simulated annual risks for composite sovereign diffusion

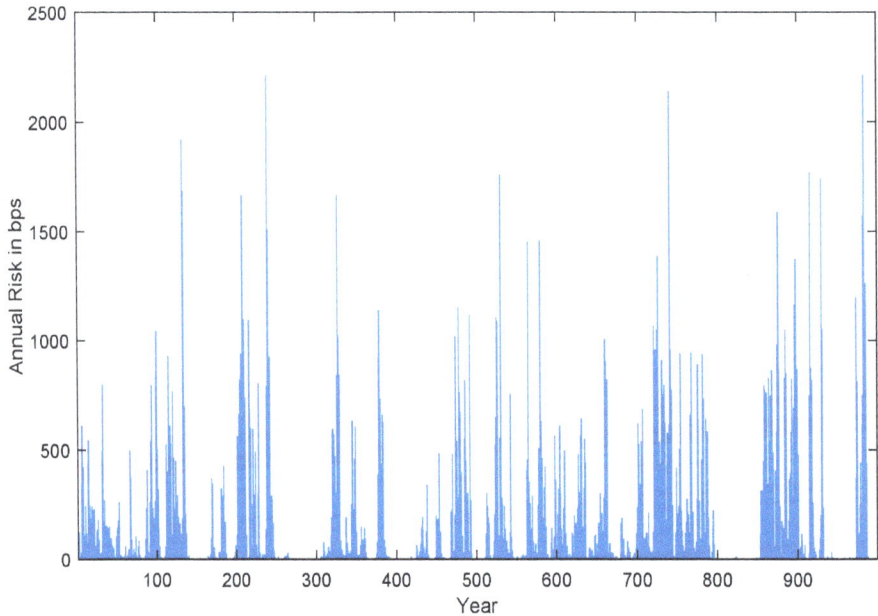

uses an EMA of 100 years duration (EMA_{100}) to make all credit forecasts and is sufficiently well-informed to identify all risks ex post. Figure 16.2 displays a 1000-year simulation, with mean and variance roughly matching the long-term norms. Absolute yearly changes average barely 4 bps and only 1% exceed 25 bps. Furthermore, the market never gets too pessimistic. Only 1% of EMA_{100} values exceed 400 bps, compared to a 15% chance of $v^* > 400$ bps and a 7% chance of $v^* > 800$ bps.

The flip side is that it is hard to get confident that risks are low. Although v^* is usually less than 50 bps, EMA_{100} dips below 50 bps less than 3% of the time. Restoration of confidence is glacial. EMA_{100} needs 35 years to halve estimated risk even when there's no trouble in the interim. Once EMA_{100} reaches 400 bps, it needs at least a century to drop below 50 bps.

Since Perfect Learner reactions depend critically on pool size N, I posit two contrasting values: $N = 80$ and $N = 1000$. The smaller N matches the sovereign benchmark. The larger N, which I consider unrealistically high, allows the inference of sovereign risks from hundreds of related assets. I call the corresponding bots PL_{low} and PL_{high} respectively.

Figure 16.2: EMA with 100-year duration for composite sovereign diffusion

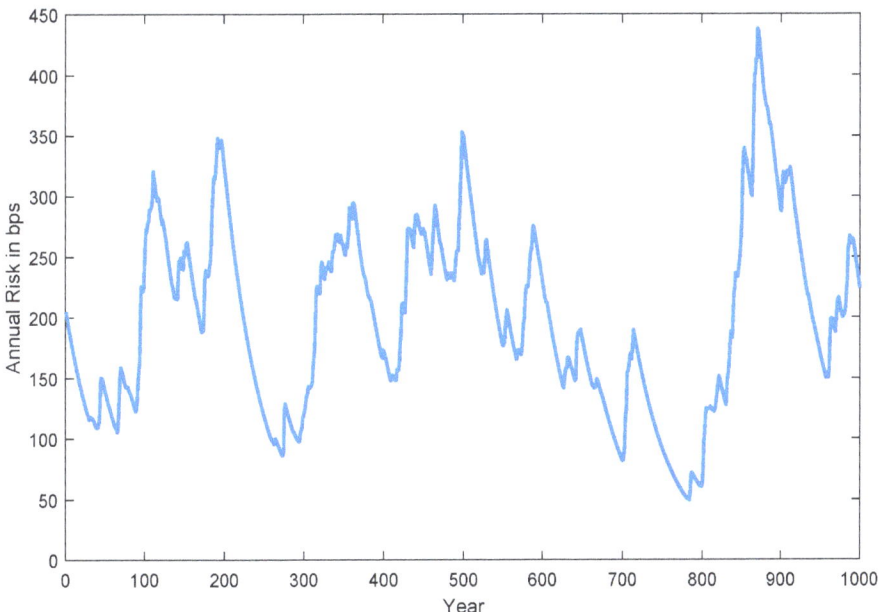

In a million-year simulation with weekly data, the median absolute deviation of E from v^* is 160 bps for EMA_{100}, 48 bps for PL_{low} and 21 bps for PL_{high}. Since E lags extreme changes in v^*, the RMSEs are much higher: 390 bps for EMA_{100}, 275 bps for PL_{low} and 182 bps for PL_{high}. For 10-year bullets, the mean absolute deviation between the perceived fair spread and the spread if v were known is 38 bps for PL_{low} and 23 bps for PL_{high}. The corresponding RMSEs are 56 bps for PL_{low} and 35 bps for PL_{high}.

Figure 16.3 compares the relative frequencies of $\log E$ and $\log v^*$. As noted earlier, EMA_{100} tracking is inherently poor when $v^* < 50$ bps or $v^* > 400$ bps. PL_{low} has trouble tracking $v^* < 5$ bps or $v^* > 2000$ bps. PL_{high} can't match $v^* < 0.3$ bp. There is a 12% chance of $v^* > 500$ bps, in which case v^* has mean 1050 bps. The corresponding conditional means of E are 237 bps for EMA_{100}, 682 bps for PL_{low} and 909 bps for PL_{high}. There is a 21% chance of $v^* < 1$ bp, in which case the conditional means of E are 150 bps for EMA_{100}, 17 bps for PL_{low} and 3 bps for PL_{high}. Not surprisingly, the absolute biases are greatest for high risk while the relative biases are greatest for low risk.

Figure 16.4 displays the relative frequencies of log spreads on 10-year bullets. Neither PL_{low} nor PL_{high} can identify high degrees of safety. The lowest spreads are 35 bps for PL_{low} and 18 bps for PL_{high}. Those levels mark respectively the 9%-ile and 6%-ile of fair spreads if v^* were known. At the high end, PL_{low} and PL_{high} are respectively 0.8% and 2.3% likely to reporting a spread over 350 bps, whereas 3.2% of v^* justify a spread that high. Hence even Perfect Learners tend to understate very high sovereign risk and overstate very low sovereign risk. The relative distortion is worst at the low end. While floors on long-term spreads often look like risk premia, they might simply reflect the challenges of identifying extreme safety.

Let us now turn from overall tracking to the dynamics. If people underestimate risks before default and drastically increase their estimate after, they will in hindsight seem unduly myopic. For PL_{low}, defaults occur on average about 1.4 weeks per year. The average v^* in those weeks is 942 bps, over 50% higher than the mean E of 600 bps. For the evidence-rich PL_{high}, defaults occur on average 10 weeks per year. The average v^* in those weeks is 656 bps, barely 10% higher than the average E of 595 bps.

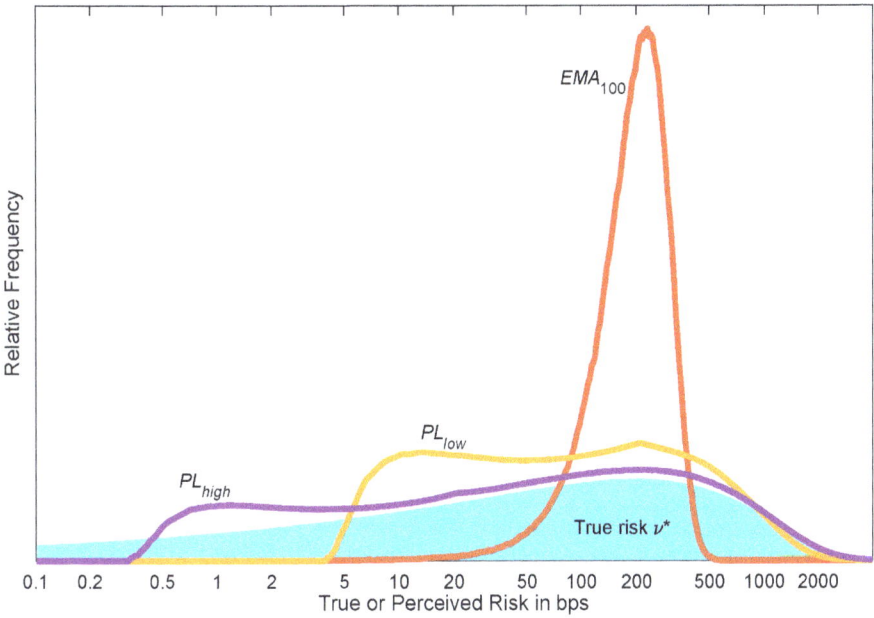

Figure 16.3: Relative frequencies of $\log E$ versus $\log \nu^*$ for EMA_{100}, PL_{low} and PL_{high}

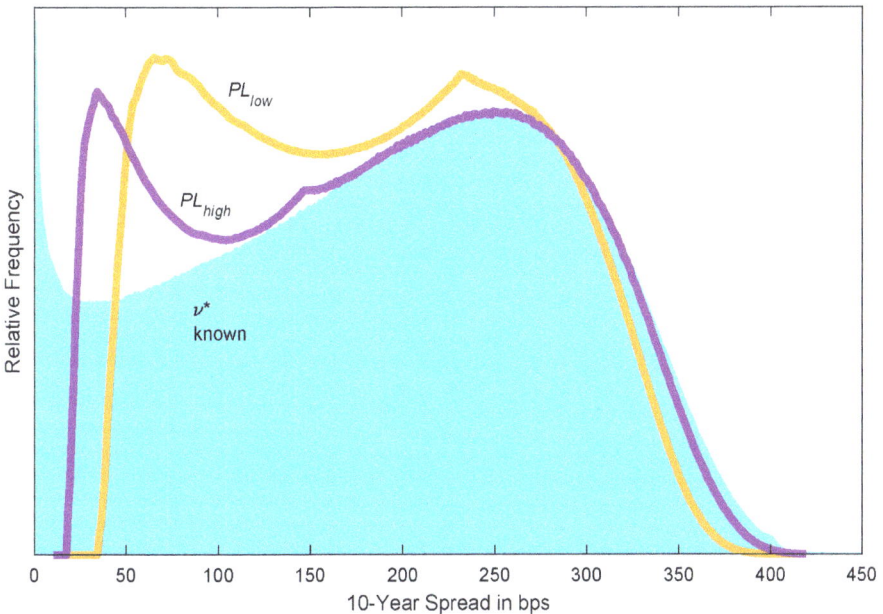

Figure 16.4: Relative frequencies of 10-year credit spreads

Hence PL_{low} tends to look extremely myopic while PL_{high} does not. Yet there is no overall bias. The understatements when defaults occur are counterbalanced by overstatements of risk when defaults don't occur.

A better measure of sensitivity to news, introduced in Chapter 12, is the effective sample duration T_{eff}/N. Figure 16.5 displays scatter plots of duration versus E for both PL_{low} and PL_{high}. Duration generally rises as E falls because observers want more defaults for identification. For similar reasons, duration is higher for PL_{low} than for PL_{high}. However, in no case does effective duration exceed 0.6 years.

For a more dramatic measure of sensitivity, Figure 16.6 compares E just after default to E just before. After default, PL_{low} raises E to at least 200 bps, no matter what it was before, as if the market were chronically prone to complacency and panic. For PL_{high} the floor is much lower, barely 20 bps, which seems more sober-minded. The spreads on 10-year bullets experience similar jumps. For PL_{low}, spreads just after default are at least 230 bps, higher than 65% of all simulated spreads. For PL_{high}, spreads just after default are at least 145 bps, higher than 35% of all simulated spreads.

Figure 16.5: Scatter plots of effective sample duration versus E

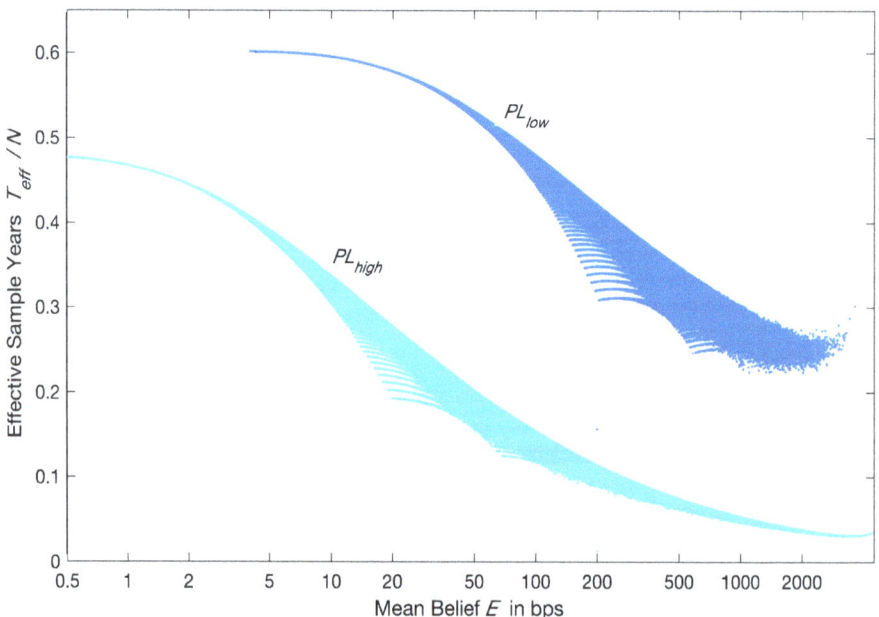

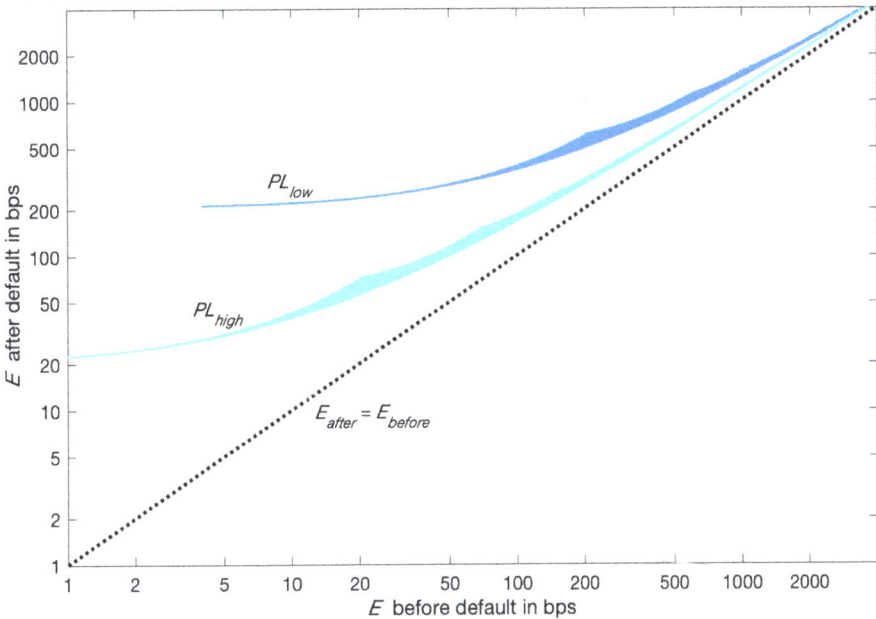

Figure 16.6: Mean beliefs E just after default compared to E just before

Perpetually Deferred Perpetuities

Our analysis demonstrates that small pool sizes and unstable risks promote a rational focus on recent data. On the one hand, this helps restore confidence after crisis. On the other hand, confidence can easily slide into complacence and flip to despair. Here are some questions that challenge this explanation along with my responses.

Q: *If sovereigns accumulate too much debt, debt markets aren't providing enough discipline. How can that be rational?*

KO: Rational doesn't mean prescient. No one can sample the future. The market can only discipline based on current evidence and current beliefs.

Q: *Surely some people smell trouble. Won't their doubts affect the market?*

KO: Of course, but as long as debts get serviced, optimistic lenders will profit and the weight of pessimists will decrease. Allowing for uncertainty about long-term E_0 would magnify the jumps in Figure 16.6.

Q: *What if sovereigns can't service without contracting more debt?*

KO: Few sovereigns repay debts without rollover. At best they outgrow their debts, so that the relative servicing burden declines.

Q: *That helps make confidence or despair self-fulfilling. Shouldn't you incorporate those effects?*

KO: Ideally, yes. A more sophisticated model might also allow for a downward trend in long-term default risk on evidence that wars and economic chaos are getting rarer.

Q: *Does that mean that this time might really be different?*

KO: That's always partly true. For example, tax systems are getting more efficient and wide-reaching, which raises sustainable debt-to-GDP ratios.

Q: *Yet spending and the costs of sovereign guarantees are growing even faster. Aren't central bankers worried?*

KO: Yes, they worry about both the borrowing spree and the crisis they might provoke by reining it in (King 2016). To delay a reckoning, they're devising indefinitely deferred perpetuities.

Q: *What does that mean?*

KO: A perpetuity pays interest regularly but never redeems the principal. Suppose the government issues new perpetuities to cover the interest. Everything is funded by the debt expansion itself.

Q: *Nearly all sovereign debt carries a fixed term. How does that resemble a perpetuity?*

KO: If fixed term debt is always repaid by issuing more debt, the aggregate still behaves like a deferred perpetuity.

Q: *Don't fixed-term investors have the option not to roll over?*

KO: Yes, in which case central banks buy the debt themselves or encourage the banks they oversee to hold it.

Capital markets will tell us when this can't work, as long-term loans will dry up and short-term interest rates soar. Yet they may offer few interim warnings.[23] As long as borrowers seem to faithfully service their debts, markets tend to get reassured, which in turn facilitates rollover and easy servicing.

The best way to lower perceived default risks is to occasionally wind down the relative debt burden. As Philip Hoffman et al (2007) have noted. Britain did this repeatedly during the 18th century, thanks partly to lender sway in Parliament. That reputation allowed Britain during the Napoleonic Wars to accumulate debt exceeding 250% of GDP without triggering a crisis. By the eve of the first World War, nearly a century later, public debt had shrunk to around 25% of GDP without default or devaluation. High confidence facilitated the borrowing that facilitated British expansion and repayment.

In contrast, French kings in the 18th century paid at least 200 bps higher interest rates than their English rivals and frequently defaulted. Defaults reflected their chronic difficulty in covering military and court expenditures out of tax revenue. In 1788 Louis XVI convened a long dormant assembly of elites to secure approval of higher taxes. Instead they insisted on constitutional reforms and hamstrung tax collection. The ensuing revolution financed itself through paper money *assignat*s that depreciated by a factor of 100 in five years. This ruined old lenders and crippled long-term French credit markets until the 1850s.

Experience suggests that debt is more symptom of fiscal problems than cause. However, debt can have a huge impact on when and how a crisis plays out. In late medieval Spain, debt helped to defer crisis for many decades. Having limited authority to impose domestic taxes, Spain's kings financed their wars through loans repaid out of silver from imperial mines in Mexico and Peru. Between mine exhaustion and silver ships lost to storms and pirates, revenues disappointed. Ten times between 1557 and 1662, Spain suspended payments and forced creditors to take long-term bonds called *juros* in partial compensation. While the forced conversions presumably raised credit spreads, they did not ruin the debt market. For decades the *juros* stayed buoyant, which allowed Spain to continue living beyond its predatory means. Eventually the decline in revenues forced default on the *juros* and wrecked Spain's financial markets.

From a modern perspective, Spain's experience looks primitive: too little information with too shallow analysis. There is however one all-too-modern aspect, namely the narcotic impact of sovereign debt. Absent clear emergency needs to fund and a clear determination to repay once the emergency passes, it fosters a borrowing addiction. With less pressure to make hard tradeoffs in resource allocation, too many groups grab too much. The next generation gets stuck with the bill, adds to it and consigns it to the generation after. The obligation to repay gets transmuted into a perceived right to keep borrowing.

In principle, any debt-to-GDP ratio seems sustainable. By extension, any one-time rise in the debt-to-GDP ratio seems sustainable too, and so does a succession of one-time rises. The larger the sovereign debt, the less any additions seem to matter, so pressures mount to bail out aggrieved subsectors and transfer more debt to the sovereign or central bank proxy.

Capital markets worry far more about liquidity than solvency. This gives borrowers substantial leeway to defer a reckoning, especially if they are large and reputable enough to appeal to a broad spectrum of lenders. However, the more accustomed society becomes to borrowing against the future. the harder it is to change course. The debt excess keeps mounting until a crisis hits that is severe enough to trigger repudiation, inflationary meltdown, or expropriation.

While our treatment of sovereign debt markets is too stylized to gauge the risks, it should make us wary of interpreting credit spreads as reliable answers. A debt burden that continually mounts as a share of GDP is like tinder piling up in a forest. Bond markets act sometimes like deer grazing in the clearings and sometimes like forest rangers scanning for fires. Rarely do they react to the tinder before it alights. Still, the tinder matters, as it can turn a controllable fire into an all-consuming conflagration.

PART IV: EQUITY MARKETS

Equity dividends can be viewed as the trail of Brownian motion. Trend drift is much harder to track than volatility. The large fluctuations over time in price-to-dividend ratios suggest that both trend growth and discounts for risk have changed markedly over time.

Equity prices tend to be much more volatile than dividends. Moreover, the excess volatility traces jittery patterns known as GARCH behavior. Both phenomena reflect the inherently volatile volatility of rational learning.

Another puzzling feature of equity prices is the high risk premium they embed. While orthodox finance traces this to anticipation of rare consumption disasters, the implied risk aversion seems implausibly high for wealthy investors. Rational gambling offers a far more reasonable explanation, as gamblers must be paid to warrant taking on market risk.

Since risk premia depend on price volatility, volatility depends on pricing and pricing depends on risk premia, valuation is challenging and prone to dispute. Traders have strong incentives to study market beliefs and trade on their momentum.

17

Dividend Growth

Equities are ownership claims on a firm's net assets. They entitle holders to dividends out of net earnings and shares of net proceeds if the firm is sold. Like bond perpetuities, equities have no explicit maturity. However, they are much harder to price than bonds as the potential payouts are far more varied and firms have considerable discretion over whether and how to make them. Correlations with other equities and consumption require extra adjustments for risk. This chapter will sketch leading pricing methodologies, note their limitations, and point to an alternative approach with more promise.

A classical valuation method discounts expected returns at an annual percentage rate R that embodies both a preference for early payment and a premium for risk. If dividends are expected to grow at mean annual percentage rate G, the NPV next year equals the dividend plus $1+G$ times the NPV today. Multiplying by the discount factor $1/(1+R)$ indicates a fair value Q for the price-to-dividend ratio of

$$Q = \frac{1+(1+G)Q}{1+R} \quad \Rightarrow \quad Q = \frac{1}{R-G}. \tag{17.1}$$

This makes equity prices highly sensitive to shifts in perceived $R-G$. At $Q=25$, which is close to the long-term average for US stocks, a 50 bps shift in $R-G$ induces a 12.5% shift in price. At $Q=50$, which is close to the average over the past 30 years, the same 50 bps shift in $R - G$ will induce a 25% shift in price.

Figure 17.1 displays price-to-dividend ratios for the composite US S&P index since its inception in 1871. The ratios are reported monthly with numerator interpolated linearly from annual dividends and denominator averaged from daily closing prices. The scaling is logarithmic to facilitate identification of the reciprocal dividend yield on the right axis.

With average G on the order of 200 bps in real inflation-adjusted terms, the implied average R is on the order of 600 bps, far higher than the real discounts on high-grade bonds. Moreover, $R - G$ looks highly unstable. Nearly 20% of annualized dividend yields fall below 200 bps or exceed 650 bps. Ten-year moving averages, which have ranged from 150 bps to nearly 600 bps, usually exceeded 400 bps before 1950 but have rarely reached 400 bps since. Closer examination suggests a long downward trend punctuated by occasional systemic scares like World Wars I and II, the Depression, and the collapse of Bretton Woods.

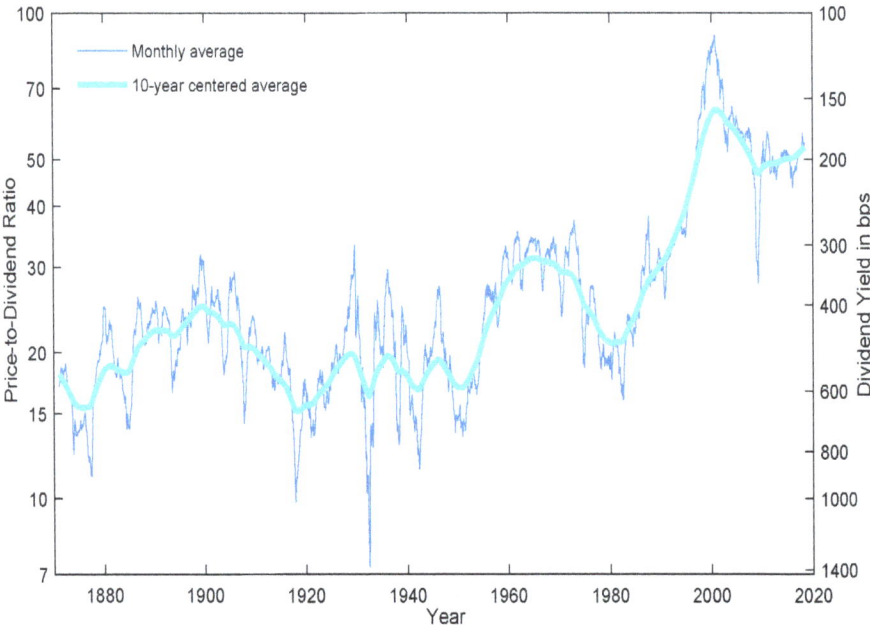

Figure 17.1: Imputed price-to-dividend ratios and dividend yields for the S&P index 1871-2018 from Shiller (2018)

What drives the changes? One candidate explanation is taxation. At first glance, Q should vary inversely with taxes to mitigate their impact on post-tax dividend yields. Unfortunately, this makes the trends even harder to explain. Q averaged much higher between 1953 and 2003 when dividends were highly taxed in the US than before 1953 when they were mostly untaxed. However, dividend taxes induced firms to repay investors in other ways, particularly through buybacks of appreciated equity shares. While most dividend taxes were cut in 2003 to 15%, many firms continue to use buybacks as dividend alternatives. As a result, some of the post-1950 rise in Q is likely spurious.

Another candidate explanation is changes in anticipated G. Judging from the cumulative moving average (CMA) of real dividend growth, it is tempting to infer a stable trend with a mean near 200 bps and a standard deviation of barely 10 bps. In support of this interpretation, the mean is broadly in line with estimates of average real GDP growth, as it should be long-term to keep the profit-to-GDP ratio from veering to extremes. However, CMA measures exaggerate convergence since their incremental weights diminish over time.

Figure 17.2 embeds the CMA in a chart of annual percentage growth. The latter is highly variable with average 11% standard deviation. If the whole series is treated as iid, the standard deviation of the sample mean is 90 bps. If only the second half of the history is treated as iid, the standard deviation is halved but so is the precision of the estimated mean, leaving a 95% confidence interval over 200 bps wide. We can trim that interval by recognizing dependence across the yearly observations. Dividend growth is positively correlated with growth the next year and negatively correlated with growth two to five years forward. The positive correlation suggests short-term smoothing of dividends to help stabilize investor returns. The negative correlation suggests longer-term reversion to trend. Roughly a third of growth surprises do not persist.

To highlight the trends, Figure 17.3 displays on a log scale an index of real dividends for the S&P. Leaving aside the impact of major wars and depressions, long-term growth can be viewed in three different ways: relatively stable, slowing mildly after 1950, or accelerating with the recent

Figure 17.2: Annual and CMA percentage growth of real dividends for the S&P 1871-2018, from Shiller (2018)

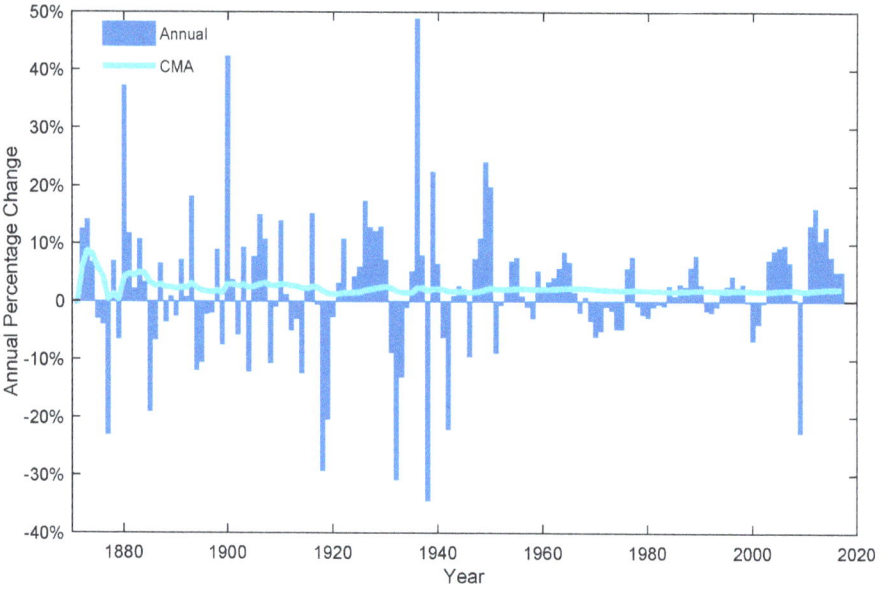

Figure 17.3: Indices of real dividends and real earnings for the S&P from Shiller (2018), initialized to one in 1871

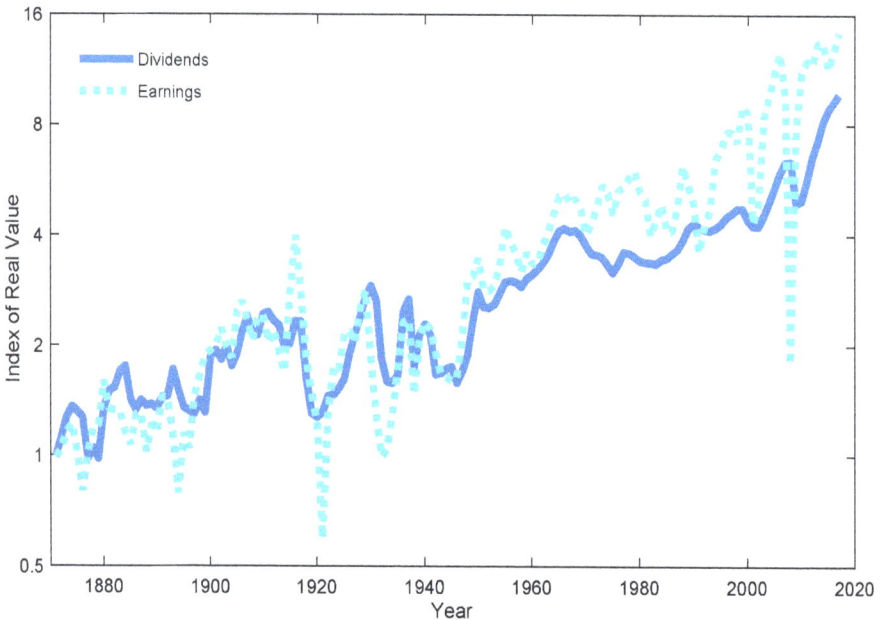

information revolution. The choice can make a 200+ bps difference in forecasts of future growth rates.

Figure 17.3 also overlays a graph of real earnings. While earnings are far more volatile than dividends, their trend growth matched that of dividends down to 1950, as it should in the long-term. Since 1950 earnings have grown slightly faster on average than dividends, with less tax-related deceleration in the 1950s or acceleration around 2000.

If growth trends haven't changed much, (17.1) will attribute most changes in Q to changes in R. Surely R has been shrinking long-term. Rising wealth encourages this, as it makes marginally deferred gratification less uncomfortable. Corporate accounts and governance are better trusted. Fears of disasters—wars, revolutions, mass expropriation, economic collapse—have also receded overall.

Contra Knight and Keynes, risks don't need to be measurable for us to quantify beliefs and test them against the evidence. The updates change our willingness to pay in predictable ways. The more uncertain we are, the more skittishly we respond to news. Where rational myopia agrees with Knight and Keynes is that the analysis of uncertain beliefs is inherently uncertain. Many market fears are prompted less by a clear surge in exogenous risk than by anxiety over its uncertain magnitude and the accompanying market turbulence.

Let σ^2 denote the variance of percentage dividend growth. For a risk-neutral observer, σ is irrelevant to valuation since (17.1) still applies to expectations. Yet σ can't be irrelevant to gamblers seeking to grow their wealth long-term. Making nothing is bound to outperform equal odds of tripling wealth or losing everything. If they aim to maximize the expected log growth g, that is usually about $\frac{1}{2}\sigma^2$ less than expected percentage growth.

We can incorporate this into (17.1) by adding $\frac{1}{2}\sigma^2$ to R or by replacing G with g. However, this compensates gamblers only for the risks of current dividend growth. It does not compensate them for residual risks in equity value, which depend mostly on beliefs about long-term trends. How should valuation adjust for them?

Perceived Growth Risks

Orthodox theorists avoid this question by positing homogeneous agents making the same forecasts. Since no one disagrees on fair prices, there are no endogenous market risks. Market risks are further reduced by identifying aggregate equity dividends with aggregate consumption. Since consumption growth is far less volatile than dividend growth—on the order of 2% annually versus the 11% in Figure 17.2—this drastically reduces the perceived σ^2. The combination didn't seem to provide nearly enough risk to justify the risk premia embedded in equity valuations.

Thomas Rietz (1988) noted that anticipation of rare disasters might resolve the puzzle and this view has gained support thanks to work by Robert Barro (2006) and others. Imagine agents with constant relative risk aversion γ and time preference ρ who face known iid risks forever, invest in either risk-free bonds or aggregate production, and seek to maximize the expected utility of aggregate consumption. Ian Martin (2013) derived a crisp solution using the cumulant generating function K of the growth Δx of log consumption. The cumulant generating function is the logarithm of the moment generating function, so here

$$K(b) \equiv \log\left\langle e^{b\Delta x} \right\rangle \equiv \log\left(\sum_{n=0}^{\infty} \left\langle (\Delta x)^n \right\rangle \frac{b^n}{n!} \right). \tag{17.2}$$

Note that $K(0)=0$ and $K(1)=\log(1+G)\equiv g$. Martin showed that the equilibrium dividend yield, which given the modeling assumptions equals the consumption-to-wealth ratio, is

$$Q^{-1} = \rho - \xi K(1-\gamma), \tag{17.3}$$

where ξ is a multiplier reflecting tradeoffs between consumption and savings.[1] Under the standard assumption of time-additive utility, $\xi=1$, in which case the dividend rate is $\rho-g$ for the risk-neutral case $\gamma=0$. When $\gamma=1$, the dividend rate is ρ regardless of growth or ξ. Neither of these expressions can account for high average dividend rates, much less their variation.

For more insight, let us examine Martin's solution for the log interest rate r_f on risk-free bonds and the extra risk premium r_p for equities:

$$
\begin{aligned}
r_f &= \rho - K(-\gamma) + (1-\xi)K(1-\gamma) \\
r_p &= g + K(-\gamma) - K(1-\gamma)
\end{aligned}
\tag{17.4}
$$

This implies $Q^{-1} = r_f + r_p - g$, the continuous-time analogue to (17.1). When consumption growth is log-normal, the risk premium is $\gamma\sigma^2$ and requires $\gamma > 10$ to fit the evidence. This is ridiculously high, as it implies no agent will pay as much as 1.2% of wealth for a 10% chance of infinite riches.

Assuming moderately high risk aversion of $\gamma \approx 4$, $K''(-\gamma)$ must be highly positive, say due to a 1%+ chance of a 30%+ plunge in consumption. Historical disaster risks are nominally consistent with that. However, roughly half of big drops are subsequently reversed, and after adjusting for this Emi Nakamura et al (2013) estimated $\gamma \approx 6$ and $\xi \approx -0.1$. Their γ estimate seems implausibly high, as it implies unwillingness to pay more than 2% for a wealth for a 10% chance of infinite riches, while their ξ estimate falls well below consensus.

Another concern is the sensitivity of r_f in (17.4). When ξ is small, r_f moves nearly as much as r_p with changes in disaster risk and γ albeit in the opposite direction. It follows that changes in perceived risk or tolerance for risk should impact r_f far more than Q^{-1}, which sums r_f and r_p. Practice belies this prediction, as inflation-adjusted rates on high-quality bonds have varied far less than the dividend yields in Figure 17.1.

My biggest concern is that equities aren't claims on all of GDP or even on a relatively stable share. Since World War II, US corporate after-tax profits have averaged less than 7% of GDP, with large cyclic variations and evidence of changing secular trends. Corporate after-tax profits outpaced GDP by an average 200+ bps per year through 1968, lagged by an average 100+ bps per year through 1992, and have outpaced GDP by an average 240 bps per year since. See Figure 17.4.

Will the information revolution make higher profit shares the new norm? Will rising public debt trigger higher corporate taxes? Will regulations become more burdensome or less? Different investors will weigh the

Figure 17.4: Corporate after-tax profits as share of quarterly US GDP 1947-2018 (US Bureau of Economic Analysis 2018)

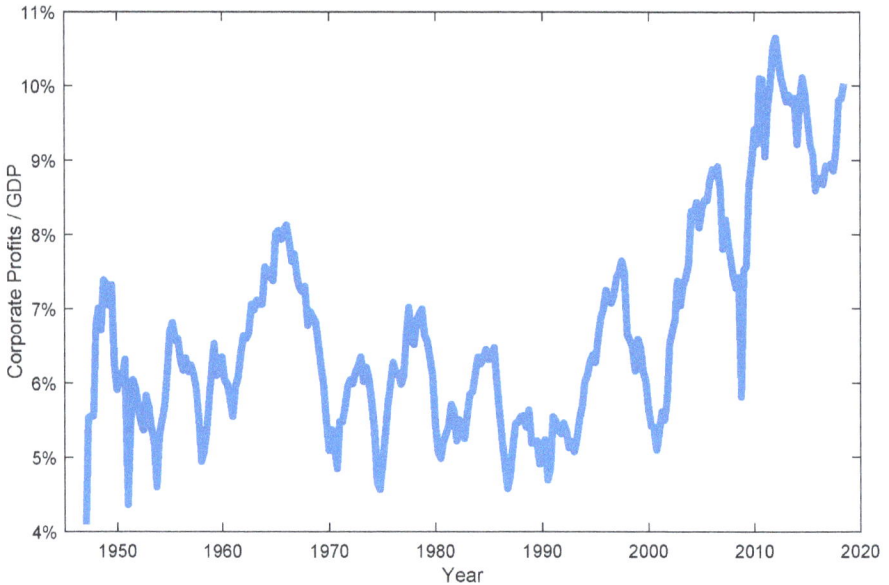

risks differently. For a stylized view, let's cultivate a bunch of gambler bots with the following agreements and disagreements:

- Bots regard any trend g between 0 and 400 bps as feasible. Prior beliefs about g have means of 100-300 bps and standard deviations of 75-115 bps. I model this using 49 beta distributions with parameters α and β spaced linearly between 1 and 3.[2]

- Bots have tiny doubts about the stability of g. They expect g to persist on average 500 to 10,000 years before resampling from the prior distribution. I model this as a yearly reset probability λ fixed at 1, 5, or 20 bps.

- Bots believe that log dividend growth is iid but are uncertain about the tail risks. I model their uncertainty as equal post-reset priors over a Student t-distribution, with 25 choices for standard deviation (linearly spaced between 3% and 20%) and four choices for degrees of freedom (5, 15, 50 and 150).[3]

- Bots observe S&P dividend growth at the end of every year and update their beliefs rationally.

Figure 17.5 displays mean log growth rates for the various bots. It takes over 100 years for the initial 200 bps range to shrink to 150 bps. The sluggishness reflects the high variance of dividend growth relative to the variance of the priors. Convergence speeds as dividend growth gets more stable and gives more evidence of a slowing trend. Disagreements swell again after the crash in 2009.

Each bot's assessment varies far more after 1960 than the CMA in Figure 17.2. This demonstrates again the huge importance of tiny doubts. The thick line is a Bayesian-updated consensus for an aggregator with equal priors over the different bots, reset at rate 0.0001 and 15 degrees of freedom for the perceived t-distribution of dividends. The reactions to the 20%+ drop in dividends in 2009 are wild. Since dividends grew slowly but relatively stably the previous three decades, most bots became convinced that volatility was very low. However, the crash evinced such high volatility that risks had seemed to reset, prompting bots to revert to their priors.

Despite the huge spread, the consensus varies barely 150 bps over the entire period. To generate more variation, we need to let g range more widely and be measured more frequently.

Figure 17.5: Estimates of g for 147 bots observing real S&P dividend growth

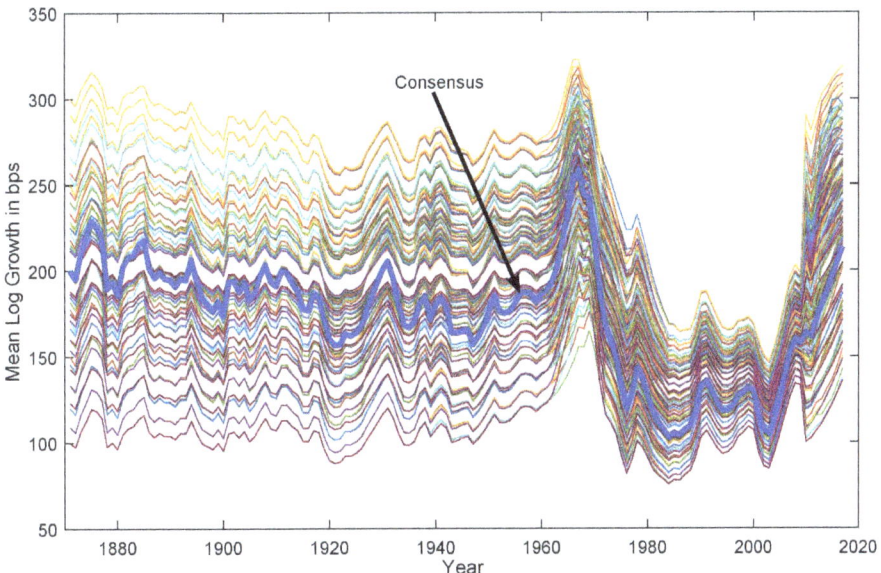

18

Brownian Bets

This chapter shifts from yearly estimation of dividend growth to continuous estimation. Specifically, I will pretend that dividends are constantly evolving diffusions, that current values are constantly and accurately reported, that these reports generate most of the news that traders draw on, and that traders evaluate news instantly. Modeling this requires tighter specification and more abstraction.

Brownian motion can be described as $dx = \mu dt + \sigma dz$, where dz denotes continual iid noise with mean 0 and variance dt. Suppose we take N measurements of interim returns Δx_i over a total time T and use them to estimate drift and volatility. No matter how frequently we measure, the best estimator $\hat{\mu}$ of a constant μ is the aggregate change divided by T and the only way to make it more precise is to extend T. In contrast, the standard deviation of the σ-estimator $\sqrt{\sum_i (\Delta x_i - \hat{\mu}\Delta t_i)^2 / T}$ approaches $\sigma^2/(2N)$. This lets us identify σ quickly if we measure frequently enough, whether or not μ is stable and whether or not we net out its estimates from the Δx_i.

The Bayesian beliefs $p(\mu)$ will capture the residual uncertainty. They should be updated in proportion to the conditional Gaussian probabilities

$$\frac{1}{\sigma\sqrt{2\pi\Delta t}}\exp\left(-\frac{(\Delta x - \mu\Delta t)^2}{2\sigma^2\Delta t}\right) \propto \exp\left(\frac{\mu}{\sigma^2}\Delta x - \frac{\mu^2}{2\sigma^2}\Delta t\right) \approx 1 + \frac{\mu}{\sigma^2}\Delta x .$$

The last step follows from a second-order Taylor expansion and heuristic approximation $(\Delta x)^2 \approx \sigma^2 \Delta t$. For the revised probabilities $p + \Delta p$ to integrate to one, $1 + (\mu/\sigma^2)\Delta x$ must be multiplied by

$$\frac{1}{1 + (E/\sigma^2)\Delta x} \approx 1 - \frac{E}{\sigma^2}(\Delta x - E\Delta t)$$

for $E \equiv \langle \mu \rangle$, which implies

$$\Delta p(\mu) \approx p(\mu)\frac{\mu - E}{\sigma^2}(\Delta x - E\Delta t).$$

Taking the limit and adjusting for anticipated shifts yields the Core Learning equation for Brownian motion:

$$dp(\mu) = p(\mu)\frac{\mu - E}{\sigma^2}(dx - Edt) + \langle \text{net migration into } \mu \rangle. \qquad (18.1)^4$$

As explained in Chapter 4, we can regard p as the fractions of total gambling capital backing various beliefs, where each gambler aims to maximize long-term wealth. For a security that costs Fdt and pays dx, a gambler anticipating a mean return μdt will strive to hold the number w of securities per unit capital that maximizes

$$\langle \log(1 + w(dx - Fdt)) \rangle \approx \langle w(dx - Fdt) - \tfrac{1}{2}(wdx)^2 \rangle .$$
$$= w(\mu - F)dt - \tfrac{1}{2}w^2\sigma^2 dt$$

Hence $w = \dfrac{\mu - F}{\sigma^2}$, consistent with (4.1). It follows that:

- The market-clearing price sets $F = E$.
- The first component of (18.1) records the redistribution of capital after dx is observed.
- The second component describes changes in relative holdings due to differential taxation or anticipated shifts in μ.

There is one small problem with this comparison. Our model x describes log dividends while the closest market counterpart trades the nominal dividends e^x. Fortunately, e^x is Brownian too with drift $\left(\mu + \tfrac{1}{2}\sigma^2\right)e^x$ and volatility σe^x. In betting on nominal dividends, the optimal Kelly bet changes but the risk exposure, dynamics and expected profit remain the same. This is easier to grasp if we rewrite the first component of (18.1) as

$$dp(\mu) = p(\mu)S(\mu)dy \quad \text{for } S(\mu) \equiv \frac{\mu - E}{\sigma} \text{ and } dy \equiv \frac{dx - E\,dt}{\sigma}. \quad (18.2)$$

$S(\mu)$ measures the Sharpe ratio of a bet when the true drift is μ while dy measures the unexpected news rescaled to standard deviations. This applies equally to dividends and log dividends.

A bigger problem with (18.1) is its demand to update p infinitely fast over countless μ. No discrete counterpart can ensure that $p + \Delta p$ stays positive. A fine grid of μ over daily Δt, coupled with either non-negativity constraints on p or use of the exponential form of h, usually provides good approximation. However, simpler solutions are preferred if we can find them. The easiest partial solution updates E using variance V, since multiplying (18.1) by μ and integrating indicates that

$$dE = \frac{V}{\sigma}dy + E^{shift}dt, \quad (18.3)$$

where E^{shift} denotes the changing drift in E expected from migration. The corresponding update for V is

$$dV = \frac{\kappa_3}{\sigma}dy - \frac{V^2}{\sigma^2}dt + V^{shift}dt, \quad (18.4)$$

where the third cumulant κ_3 equals skewness times $V^{3/2}$ and where V^{shift} denotes the changing drift in variance expected from migration.

If skewness is permanently zero, we can stop there: beliefs stay normally distributed and all we need to track is their mean and variance. Such learners, known as Kalman filters, work well when the core drivers of μ are well understood, all change is gradual and noise is symmetric around the mean. They are optimal when μ is a diffusion.

If μ is anticipated to follow Brownian motion with zero drift and volatility δ, the Kalman filter boils down to $dE = (V/\sigma)dy$ and $dV = (\delta^2 - V^2/\sigma^2)dt$ as suggested in (8.1). When δ and σ are stable, this converges to $dE = \delta dy$, which is an ordinary EMA with learning rate δ/σ and makes E just as volatile as μ. Since the implied long-term $V = \delta\sigma$, the standard deviation $\sqrt{V}$ of beliefs equals the geometric mean of δ and σ, which matches the long-term RMSE. For example, if $\sigma = 1000$ bps per year and $\delta = 10$ bps per year, the long-term learning rate is 0.01 per year with tracking comparable to a rolling 200-year average. Yet the long-term RMSE is 100 bps, much higher than we would prefer.

Figure 18.1 displays a 200-year sample path for μ starting at $\mu_0 = 200$ bps and 20 sample paths for E assuming $E_0 = \mu_0$ and $\sqrt{V_0} = 150$ bps, with all values updated daily for 250 trading days per year. Each E path responds to a different pattern of $\sigma\,dz$ noise, which isn't displayed. The common feature is that none of the E track μ very well. They can't. The noise is too intense to muffle without centuries of averaging, yet δ degrades the relevance of centuries-old observations. The only reason individual E paths don't range more widely is that each path updates the mean for a single sample history. The median absolute change over 50 years is 50 bps, which is only half the long-term RMSE.

Between the relatively stable V and the moderation in E, this model cannot capture the turbulence of equity markets. To modify it, let's drop the assumption that observers know the precise drivers of μ. Also, instead of applying the same Kalman filter to different observations, let's look at different Kalman filters applied to the same observations. Suppose these filters choose among $\{100,200,300\}$ bps for E_0, $\{5,10,20\}$ bps for δ, and $\{-1.5,0,1.5\}$ bps for E^{shift}. Suppose also that V_0 matches long-term V, which reduces the Kalman filters to EMAs with drift. Figure 18.2 depicts the varied interpretations of a random 200-year stream of consumption.

Overlaid on Figure 18.2 are plots of actual μ and market price. The market price is calculated using a two-layer inter-threaded market with reset rates ranging from 0.4 per year to 0.4 per million years. It wobbles significantly more than the actual drift or the base-level Kalman filters.

Figure 18.1: Kalman filters for 20 different sample paths with daily updates for 200 years given annual $(\sigma, \delta, \mu_0, E_o, \sqrt{V_o}) = (1000, 10, 200, 200, 150)$ bps

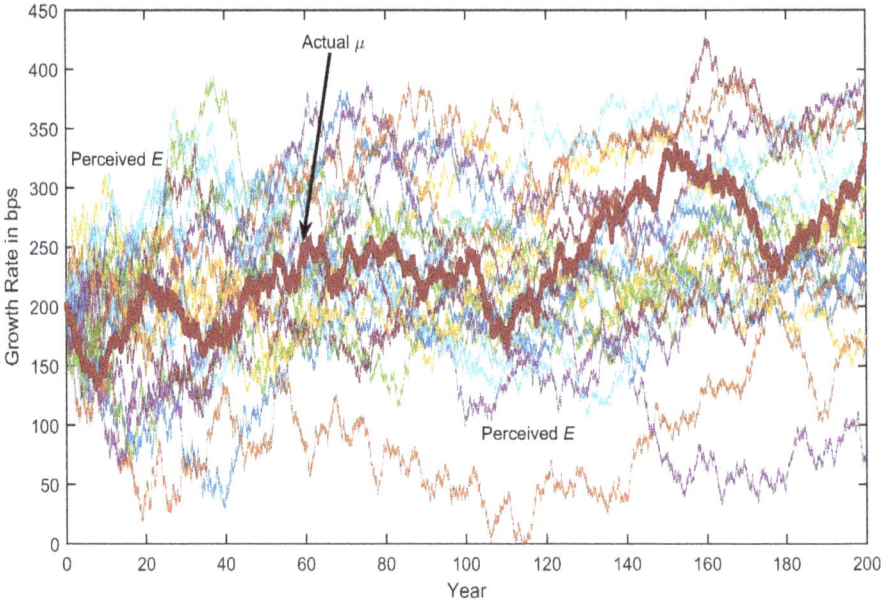

Figure 18.2: Kalman filters for 200-year sample generated with same risk drivers as in Figure 18.1, overlaid with drift and market price.

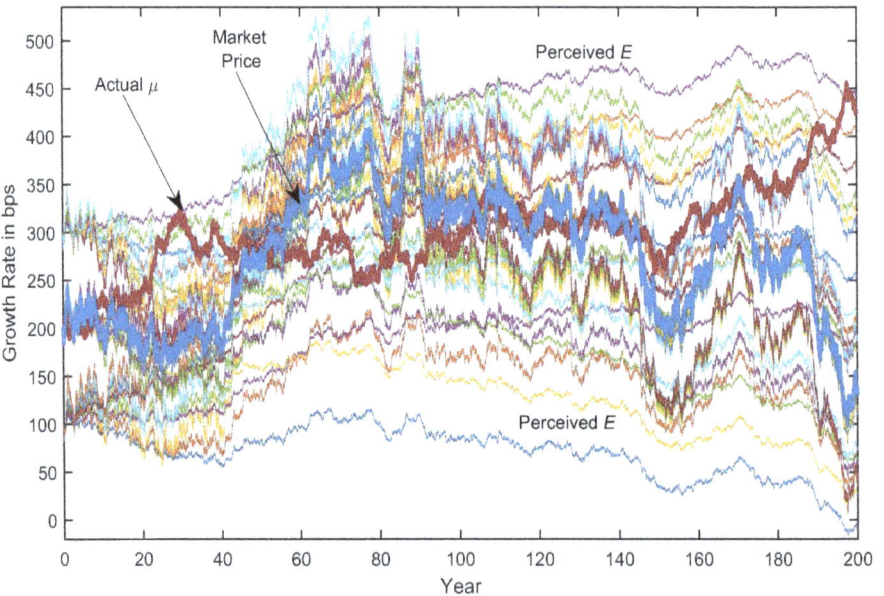

When the sample is extended to 20,000 years, the market price has only 5% higher RMSE than the optimal filter but over twice the volatility. The extra volatility has two sources. The first source is disagreement, as the aggregate variance of beliefs exceeds the mean individual variance by the variance of the individual means. The second source is reweighting of individuals' credibility through Bayesian updates.

The extra price volatility in Figure 18.2 doesn't change much over time because fundamental change stays modest relative to the noise. Relative skewness κ_3/σ in (18.4) rarely gets large enough for long enough to drive big changes in V.

In contrast, V is inherently unstable when μ is viewed as prone to large shocks. Suppose μ equals 300 bps less a gamma disturbance of shape 0.5 and mean 100 bps that resets on average once every 25 years. Suppose market participants understand the general structure but disagree whether the long-term mean is 100, 200, or 300 bps, whether the shape is 0.5 or 1, whether the gamma mean is 50 or 100 bps and whether the yearly reset rate is 0.02 or 0.04. Figure 18.3 displays a simulated 200-year history with $\sigma = 0.1$ as before.

Compared to Figure 18.2, the sample histories are broadly similar in RMSE, maximum variations and average volatilities. However, Figure 18.3 exhibits far more uneven volatility. To show this more clearly, it includes a plot of the smoothed learning rate η, computed as the mean absolute price change over the past 200 days divided by the corresponding mean absolute observation. There are multiple instances where the smoothed η doubles or halves in five years or less with no change in μ. The maximum smoothed η is eight times the minimum.

The connections between μ and smoothed η are weak. While trend dividend growth in Figure 18.3 looks flat or mildly declining, the market seems overcome by what Keynes called animal spirits and Shiller called irrational exuberance. The true disruptor is rational turbulence. The top half of Figure 18.4 displays 1000 sample years of daily unsmoothed η for a Perfect Learner. There is enormous variation in the scale and duration of the surges in uncertainty. The bottom half of Figure 18.4 displays a 20-year subsample, which reveals substantial variation on shorter time scales.

Figure 18.3: Mean beliefs for 200-year sample with imperfect anticipation of gamma shocks, overlaid with drift, market price and smoothed learning rate

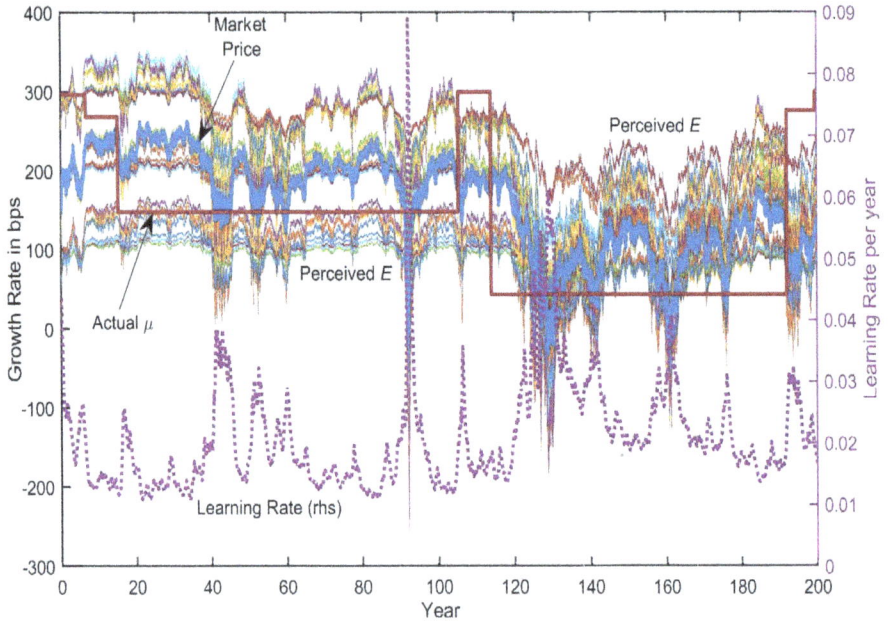

Turbulence is never completely regular. For example, while $\eta > 0.05$ typically recedes quickly in Figure 18.4, near the end of the sample it persists for most of a century. Still, there are many near-regularities. Figure 18.5 shows that $\log \eta$ is roughly normal in distribution with a mild skew while the changes $\Delta \log \eta$ approach what is known as a variance-gamma distribution, with a steep peak near zero and very fat tails.[5]

Dynamics of Turbulence

For Kalman filters, beliefs stay normal with all higher cumulants zero. For other distributions of beliefs, the cumulant hierarchy never ends. All non-normal distributions potentially breed turbulence. However, they might be close enough to normal that turbulence isn't important, like the market of EMAs tracking drift in Figure 18.2. For a mathematical demonstration, let's try to update the nth moment of beliefs. Multiplying (18.2) by μ^n and integrating, we see that

Figure 18.4: Learning rates over 1000-year sample and 20-year subsample
for Perfect Learners expecting gamma shocks

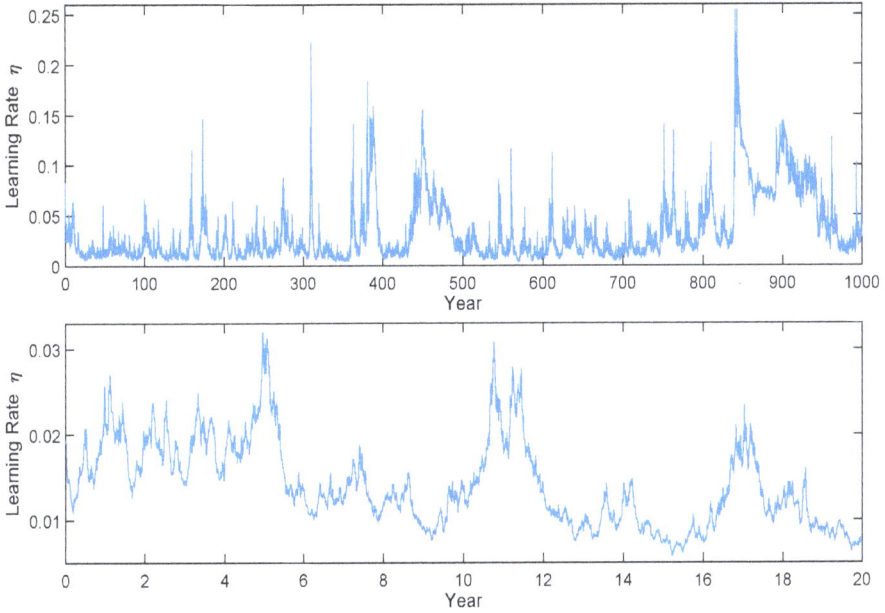

Figure 18.5: Histograms of $\log \eta$ and $\Delta \log \eta$ for Figure 18.4

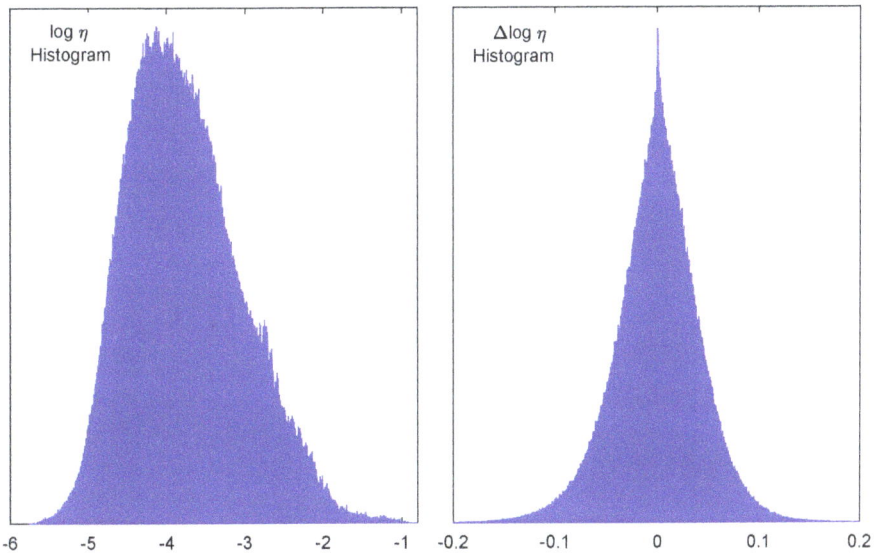

$$d\langle \mu^n \rangle = \frac{\langle \mu^{n+1} \rangle - \langle \mu^n \rangle E}{\sigma} dy \, .$$

Each moment update depends on the moment one order higher. Using the moment generating function $M(b) \equiv \langle e^{b\mu} \rangle$, or equivalently multiplying (19.2) by $e^{b\mu}$ and integrating, lets us summarize this as

$$dM(b) = \frac{\langle \mu e^{b\mu} \rangle - \langle e^{b\mu} \rangle E}{\sigma} dy = \frac{M'(b) - M(b) \cdot E}{\sigma} dy \, .$$

Conversion to cumulants is easiest if we invoke the cumulant generating function $K(b)$ introduced in (17.2). Applying stochastic calculus,

$$K \equiv \log M \quad \text{has volatility} \quad \frac{|K' - E|}{\sigma} \quad \text{and drift} \quad -\frac{(K' - E)^2}{2\sigma^2} \, .$$

Adding the cumulant generating function K^{shift} of rates of expected shifts in μ yields the update

$$dK(b) = \left(K^{shift}(b) - \frac{(K'(b) - E)^2}{2\sigma^2} \right) dt + \frac{K'(b) - E}{\sigma} dy \, .$$

The cumulants κ_n introduced in Chapter 3 are the Taylor series coefficients of K. Equating the Taylor expansions, the shifts work out to

$$d\kappa_n = \left(\kappa_n^{shift} - \frac{1}{2\sigma^2} \sum_{j=1}^{n-1} \binom{n}{j} \kappa_{j+1} \kappa_{n-j+1} \right) dt + \frac{\kappa_{n+1}}{\sigma} dy \, , \tag{18.5}$$

which match (18.3) for $n=1$ and (18.4) for $n=2$.

While the drift terms get progressively more complicated as n increases, they tend to shrink κ_n absent shifts in the underlying μ. Never do they invoke a higher-order cumulant to explain the drift of a lower-order cumulant. However, κ_n has volatility $|\kappa_{n+1}|/\sigma$ as noted in (3.3), which gives outliers enormous influence.

Imagine we have nearly Gaussian beliefs when evidence arrives of a $\langle dy \rangle \approx \chi \gg 1$ standard deviation outlier. Even if the higher cumulants are initially miniscule, each κ_n will tend to scale with χ^n. Any κ_n that gets large will reverberate on lower cumulants via (18.5). The impact can build up fast. A surge in κ_2 shows up in high volatility of price; a surge in κ_3 shows up in high volatility of volatility and so on. However, the buildups in lower cumulants tend to shrink higher κ_n via drift, while the mean dy decreases as $E = \kappa_1$ shifts towards the outlier. Turbulence will decay absent new shifts, with higher cumulants shrinking faster than lower cumulants. Once E is identified correctly, the negative drift tends to dominate and drive all higher cumulants toward zero.

The cumulant hierarchy explains the apparent bifurcation of learning. Often a calm phase dominates with slow learning and tiny higher cumulants. Sometimes a turbulent phase dominates with much faster learning and more extreme cumulants.. Transitions between these two phases tend to be quick, especially from calm to turbulence. The durations of each phase vary with the scale and timing of surprise.

For a stylized example, suppose that $\sigma = 1000$ bps, that initial beliefs are Gaussian with $E_0 = 0$ and $\sqrt{V_0} = 10$ bps, that there is a perceived $\lambda = 10^{-6}$ chance of reset from a Gaussian distribution with mean zero and standard deviation 100 bps, and that there is a surprise selection $\mu_0 = 100$ bps.[6] Figure 18.6 charts mean, variance, skewness and kurtosis during a 5000-play simulation. E changes slowly for the first 1250 observations and very rapidly for the next 900 observations. The shift is partly preceded by a huge surge in V, which recedes before E stabilizes. Skewness surges before V does and recedes as V peaks. Kurtosis surges before skewness does and recedes as skewness peaks.

To check how representative this is, Figure 18.7 displays the median mean, variance, skewness and kurtosis for 500 simulations, with each plot rescaled as a fraction of the maximum observed median. Each cumulant's surge and recession is preceded by the surge and recession of higher-order cumulants. A chart of the means looks very similar except that the core transitions are more drawn out, recalling the difference between mean and median prediction in Figure 5.3.

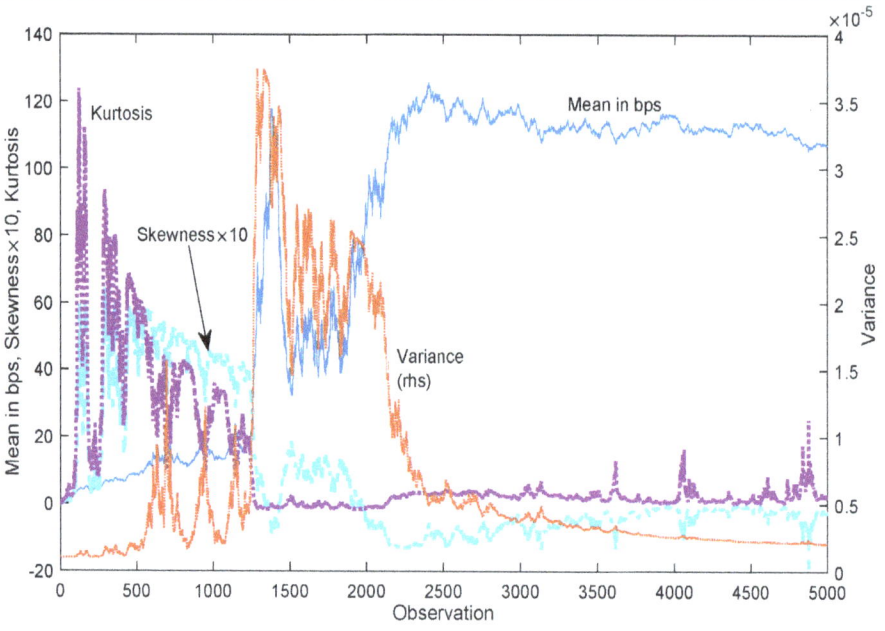

Figure 18.6: Mean, variance, skewness and kurtosis for surprise shift described in text

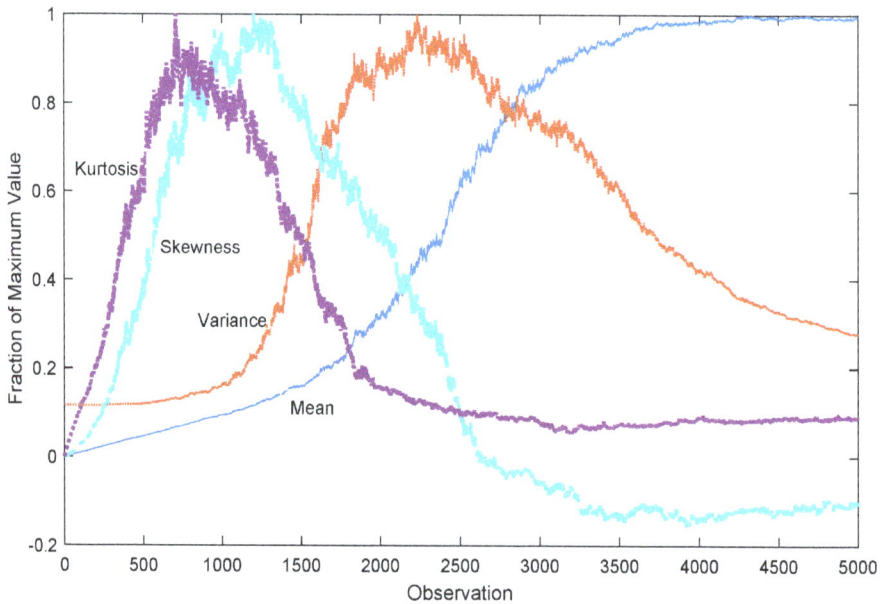

Figure 18.7: Median mean, variance, skewness and kurtosis for 500 simulations of Figure 18.6, plotted as fraction of maximum values

19

Excess Volatility

Simple NPV models with constant drift suggest that price volatility should basically match dividend volatility. It doesn't. Instead, market prices tend to be significantly more volatile than the dividends they discount. Furthermore, volatility isn't stable; it alternates between calm periods and storms of varying duration and intensity.

These behaviors are often viewed, even by experts, as market defects or symptoms of the madness of crowds. Robert Engle, co-winner of the 2003 Nobel Prize, developed GARCH models of volatile volatility that describe it neatly but have no apparent rationale. Robert Shiller, co-winner of the 2013 Prize, attributed excess volatility to irrational exuberance. This leaves many mysteries unresolved. For example, if neither excess volatility nor GARCH behavior is rationally justified, why do we rarely see one without the other?

In contrast, rational learning provides a unified explanation. Let $\overline{Q}_\mu$ denote the equilibrium price-to-dividend ratio given a known current drift μ of log dividends x. Suppose that the market consensus Q stays close to the expectation $Q_p \equiv \langle \overline{Q}_\mu \rangle$ given beliefs p. Suppose also that future drift is positively correlated with current drift, as is true for any mix of reset and diffusion. Then dQ_p will be positively correlated with dx and its standardized form dy, since (18.1), (18.2), and a Taylor series expansion of $\overline{Q}_\mu$ around $\overline{Q}_E$ indicate that

$$dQ_p = \int \bar{Q}_\mu dp(\mu) d\mu = \left(\int \bar{Q}_\mu \frac{\mu - E}{\sigma} p(\mu) d\mu \right) dy + \left\langle Q_p \text{ migration} \right\rangle$$

$$\approx \bar{Q}'_E \frac{V}{\sigma} dy + \left\langle Q_p \text{ migration rate} \right\rangle dt$$

If only discrete drifts are feasible, we can proxy the derivative $\bar{Q}'_E$ with the slope of a connecting spline. It follows that market price $P = Qe^x$ will be systematically more volatile than dividends. Extension to higher cumulants demonstrates the volatility of volatility.

In other words, the higher dividends are, the higher the likelihood that growth has accelerated and hence the higher the consensus price-to-dividend ratio. The positive correlation makes the price wobble relatively more than dividends do. That accounts for excess volatility. The volatility of that excess volatility reflects rational turbulence.

The simplest non-trivial case posits two choices $\mu_H > \mu_L$ and reset at rate λ to a stationary distribution with probability $\pi_H = 1 - \pi_L$ of high drift. Since drift can reset to itself, this is equivalent to Markov switching at rate $\lambda \pi_L$ from high to low and rate $\lambda \pi_H$ from low to high. The reciprocal of the switching rates gives the average duration T_H or T_L of each state.

When drift equals μ_H, a fairly priced investment of Q_1 pays dividends at rate 1 and grows at expected percentage rate $\mu_H + \frac{1}{2} \sigma^2$.[7] In an instant dt, there is probability $\lambda \pi_L dt$ of switching to low drift, for which the fair price-to-dividend ratio is Q_0. Applying the discount $1 - rdt$ on future returns, equilibrium prices satisfy

$$Q_1 \approx 1 \cdot dt + (1 - rdt)(1 + \mu_H dt + \tfrac{1}{2} \sigma^2 dt)\left[(1 - \lambda \pi_L dt)Q_1 + \lambda \pi_L dt \cdot Q_0 \right].$$

First-order conditions require $(\lambda \pi_L + J_H)Q_1 = 1 + \lambda \pi_L Q_0$, where $J_H \equiv r - \frac{1}{2}\sigma^2 - \mu_H$ denotes the "pure" dividend rate absent switching, analogous to $R - G$ in (17.1). When drift equals μ_L, a parallel argument requires $(\lambda \pi_H + J_L)Q_0 = 1 + \lambda \pi_H Q_1$ for $J_L \equiv r - \frac{1}{2}\sigma^2 - \mu_L$. The solution is

$$Q_p = \frac{\lambda + J_H + p(J_L - J_H)}{\lambda \langle J \rangle_\pi + J_H J_L} \quad \text{for} \quad p \equiv p(\mu_H), \tag{19.1}$$

provided drifts are low enough to keep values positive; otherwise Q is infinite. Assuming perfect learning, (18.1) simplifies for two drifts to

$$dp = p(1-p)S_{MAX}dy + \lambda(\pi_H - p)dt , \qquad (19.2)$$

where $S_{MAX} \equiv (\mu_H - \mu_L)/\sigma$ denotes the maximum possible Sharpe ratio. Since dQ is instantaneously perfectly correlated with dx, the volatility ς_p of market price satisfies

$$\varsigma_p \equiv \text{vol}(\log P_p) = \sigma + \text{vol}(\log Q_p) . \qquad (19.3)$$

By Itô's rule, the last term equals $(\log Q_p)'$ times the volatility of p. Applying (19.1)-(19.3),

$$\frac{\varsigma_p}{\sigma} = \frac{p(1-p)S_{MAX}^2}{\lambda + J_H + p(J_L - J_H)} + 1 . \qquad (19.4)$$

It follows that:

- Prices will be more volatile than dividends unless drift is effectively constant via $p=0$, $p=1$, $S_{MAX}=0$, or $\lambda = \infty$.
- Otherwise prices are more volatile than dividends.
- Excess volatility is volatile since it varies with a volatile p.
- Price volatility will vary a lot if S_{MAX} is high, λ is low, and p swings between near-certainty and uncertainty.
- Conversely, volatility will be relatively stable if S_{MAX} is low, λ is high, or p rarely goes to extremes.
- Low r raises excess volatility while high λ dampens its impact.

Let's look at five examples. The first sets $r = 0.05$, $\sigma = 0.1$, $\mu_H = 0.04$, $\mu_L = -0.01$, $\lambda = 0.1$ and $\pi_H = 0.5$. Here high and low drifts are equally likely with a gap of half a yearly standard deviation between them and an average 20 years between switches. Figure 19.1 displays Q and ς for a 100-year simulation of 250 trading days each. In a 100,000-year simulation, Q ranges from 32.5 to 46.9 with a mean of 39.7. The excess volatility averages

Figure 19.1: Fair price-to-dividend ratios Q and price volatility ς for 100-year simulation given $(r,\sigma,\mu_H,\mu_L,\lambda,\pi_H)=(0.05,0.1,0.04,-0.01,0.1,0.5)$

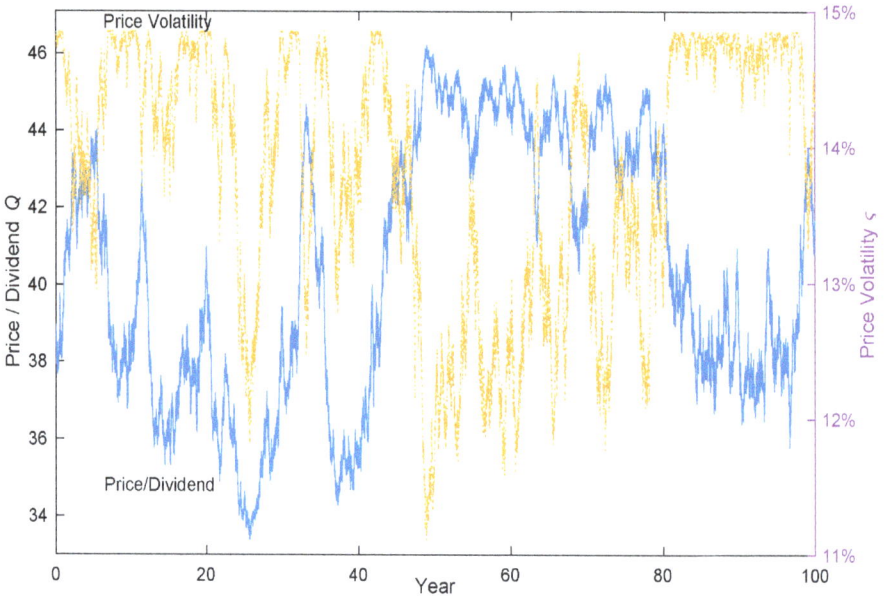

416 bps. As a result, nearly 20% of prices appreciate or depreciate by a net 20% in one year, whereas only 8% would if Q stayed constant. While this seems moderately market-like, ς is confined to an unrealistically narrow range: it never reaches 15% or dips below 10.5%.

Relative excess volatility is typically capped somewhere near $S_{max}^2/4\lambda$. To raise the cap and induce more variation, Example 2 changes σ, μ_H, and μ_L to 0.08, 0.05, and -0.03 respectively while keeping r, λ and π_H the same. As Figure 19.2 shows, the revisions expand the range of Q by over 60% and nearly triple the range of ς. Calm and stormy periods are more clearly distinguished and there are more false starts.

The simulations look more realistic apart from one strange new feature: a relatively stable equilibrium near each extreme. In a 100,000-year simulation, Q ranges from 28.3 to 51.3 with a mean of 39.8. Curiously, the mean is marginally higher than in Example 1 despite the 50 bps drop in mean μ. In this case Q_1 gains slightly more from higher μ_H than Q_0 loses from lower μ_L. ς ranges from 8.3% to 23.0% with mean 16.1%, for an average excess of 809 bps.

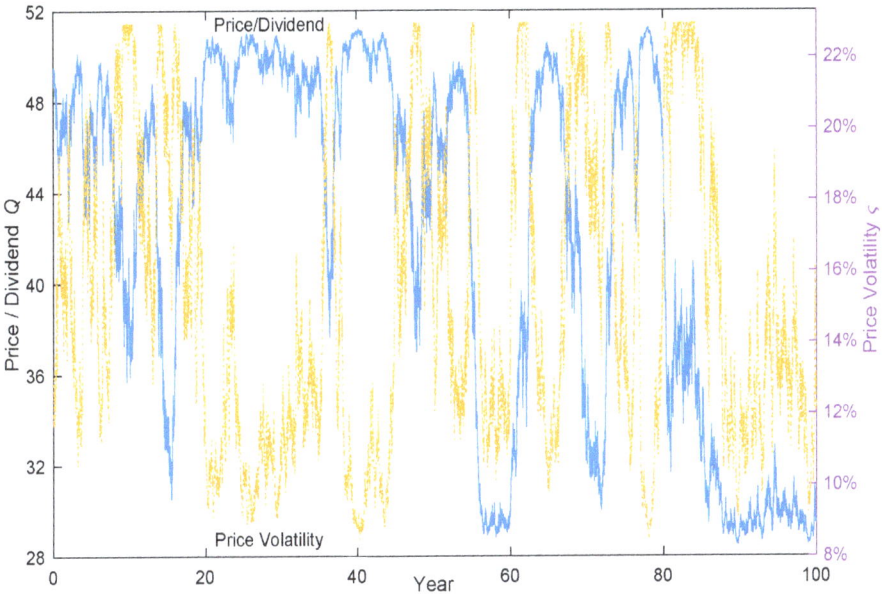

Figure 19.2: Q and ς for 100-year simulation given
$(r,\sigma,\mu_H,\mu_L,\lambda,\pi_H)=(0.05,0.08,0.05,-0.03,0.1,0.5)$

Lower price volatility for extreme values of Q is inevitable in a two-drift model, since $p(1-p)$ in (19.4) will be tiny. High S_{MAX} just accentuates the contrasts. However, the two equilibria won't be equally important unless $\pi_H \approx 0.5$. To illustrate this, Example 3 sets new values $\mu_H = 0.04$, $\mu_L = -0.13$, $\lambda = 0.24$ and $\pi_H = 0.833$. The λ and π_H choices make high drift persist an average 25 years before switching to low drift, which in turn persists an average 5 years before switching back. Figure 19.3 depicts a 100-year simulation. The baseline equilibrium pegs Q near 42 and ς near 9%. Yet the market is prone to what look like anxiety attacks. In minor scares, ς surges to 15%-20% and Q falls a few points. Major panics trigger depressions where Q falls below 30 and ς subsides in despair.

Of course, our Perfect Learner bots aren't capable of panic. They rationally expect dividends in Example 3 to nearly halve during a low-drift period of average duration and usually identify low drift confidently within a year of its start. The flip side is frequent miscues and corrections. In a 100,000-year simulation, Q ranges from 25.7 to 43.2 with a mean of 40.3. The mean far exceeds the median because two-thirds of Q values exceed

Figure 19.3: Q and ς for 100-year simulation given
$(r, \sigma, \mu_H, \mu_L, \lambda, \pi_H) = (0.05, 0.08, 0.04, -0.13, 0.24, 0.833)$

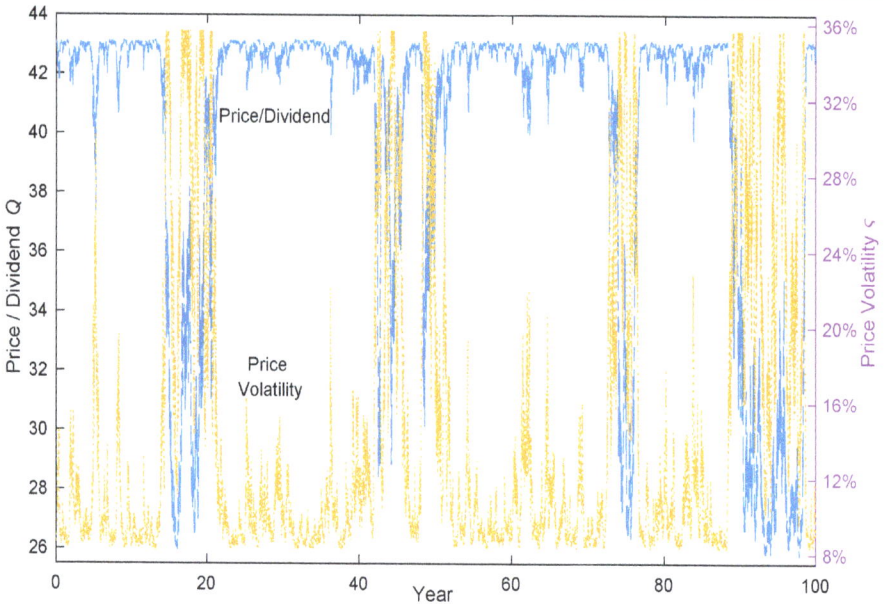

42. Price volatility ranges from 8.1% to 35.7% with mean 14.3%, for an average excess of 629 bps. Excess volatility averages 462 bps even when the underlying drift is high.

Example 4 retains r, σ, and μ_H from Example 3 but sets $\mu_L = -0.26$, $\lambda = 0.4$, and $\pi_H = 0.9$. While this shortens low-drift periods to an average 2.8 years, it makes them twice as harsh. The changes maintain the same mean for Q and nearly the same range of Q, mean dividend reduction in crisis and mean ς.[8] The simulations plotted in Figure 19.4 resemble those in Figure 19.3. Yet the volatility scale is quite different. Volatility peaks at 59.5%, nearly 2400 bps higher than for Example 3.

Example 5 makes low drift the norm rather than the exception. It sets $\lambda = 0.2$ and $\pi_H = 0.167$, which implies 6-year average duration for high drift and 30-year average duration for low drift. It also sets $\mu_H = 0.1$ and $\mu_L = -0.005$. In a 100,000-year simulation, Q ranges from 35.9 to 61.1 with a mean of 40.1, while ς ranges from 8.2% to 25.6% with mean 14.7%. Figure 19.5 displays a 100-year slice. The main novelty is that ς moves more in tandem with Q than counter to it.

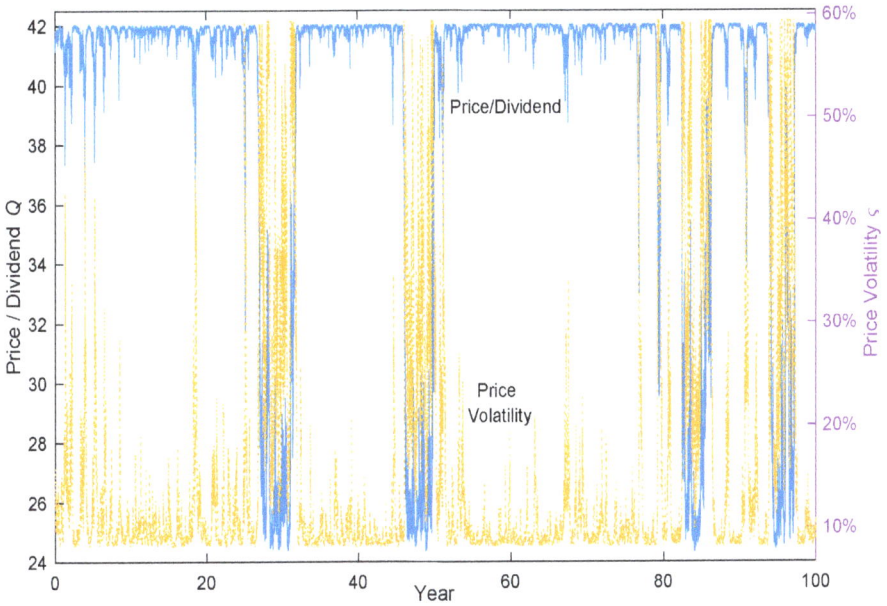

Figure 19.4: Q and ς for 100-year simulation given
$(r, \sigma, \mu_H, \mu_L, \lambda, \pi_H) = (0.05, 0.08, 0.04, -0.26, 0.2, 0.9)$

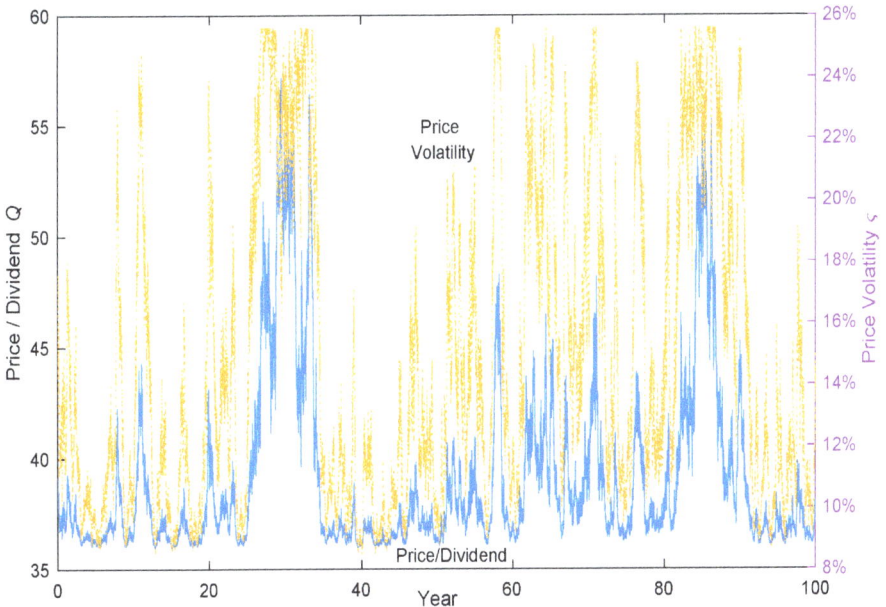

Figure 19.5: Q and ς for 100-year simulation given
$(r, \sigma, \mu_H, \mu_L, \lambda, \pi_H) = (0.05, 0.08, 0.1, -0.005, 0.2, 0.167)$

Decoding Beliefs

The characteristic features of Q and ς depend crucially on the long-term density h^* of p. The Kolmogorov forward equation implies

$$h^*(p) \propto p^{\frac{\pi_H - \pi_L}{U} - 2}(1-p)^{\frac{\pi_L - \pi_H}{U} - 2}\, e^{-\left(\frac{\pi_H}{Up} + \frac{\pi_L}{U(1-p)}\right)}, \qquad (19.5)^9$$

where $U \equiv S_{MAX}^2/(2\lambda)$. A ratio similar to U arose in analysis of (19.5), so let's think about what it represents. Suppose the market always deems both drifts as equally likely. A gambler who knows the current drift and bets full Kelly until the drift resets earns an expected profit of $\frac{1}{2}U$. This makes U a measure of potential information in the system.

   When $\pi_H = \pi_L = 0.5$, (19.5) reduces to a multiple of $z^{-2}e^{-1/(2Uz)}$ for $z \equiv p(1-p)$ and is symmetric around $p = 0.5$. This increases with z up to $1/(4U)$ but z is capped at 0.25. As a consequence, h^* is single-peaked for $U \leq 1$ and bimodal for $U > 1$. Figure 19.6 displays h^* for five different values of U given $\pi_H = 0.5$. The flattened dome is the density for the baseline $U = 1$. The solid lines outline h^* for $U = 1.25$ from Example 1 and $U = 5$ from Example 2. The dotted lines show the impact of raising U to 8 or lowering it to 0.2.

   Volatility depends crucially on z, which was charted in Figure 5.1. For p between 0.3 and 0.7, z varies by less than 20%. Hence volatility cannot change much unless $U \gg 1$ or one drift dominates. Contrasts between calm and turbulence are sharpest when both conditions hold, like in Examples 3 through 5 where $U \approx \{9.4, 17.6, 4.3\}$ and $\pi_H \approx \{0.83, 0.9, 0.17\}$. The combination makes the favored peak so tall that the stationary densities in Figure 19.7 need to be plotted in log terms to make the other peak visible.

   The more common drift is usually better identified than the rarer drift as it is more anticipated and persist longer. When drift in Examples 3 or 4 takes its usual high value, confidence p averages 93% or 97% respectively. In contrast, when drift in Examples 3 or 4 is low, confidence $1 - p$ that it is low averages 36% or 30% respectively. In Example 5, confidence averages 90% when drift takes its usual low value and 49% when drift is high.

Figure 19.6: Stationary densities for $\pi_H = 0.5$ and assorted U

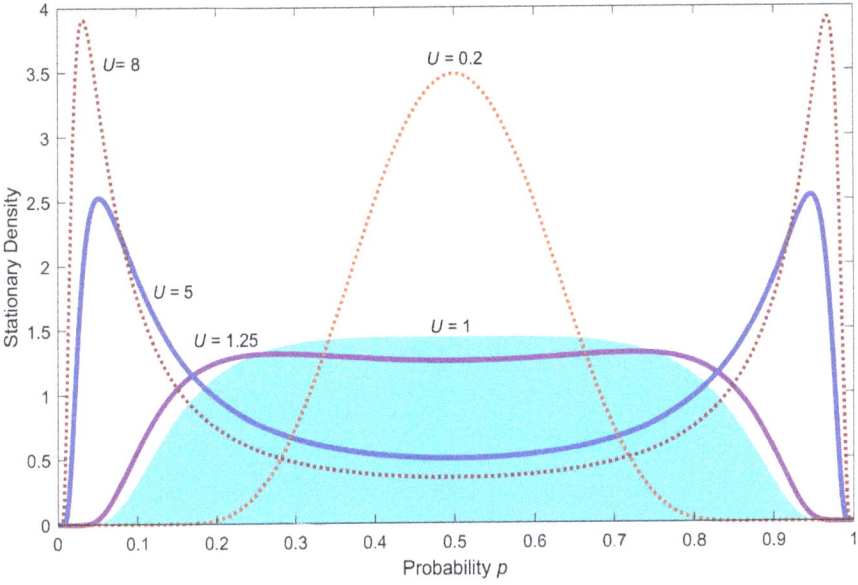

Figure 19.7: Stationary densities for Examples 3-5, plotted in log terms

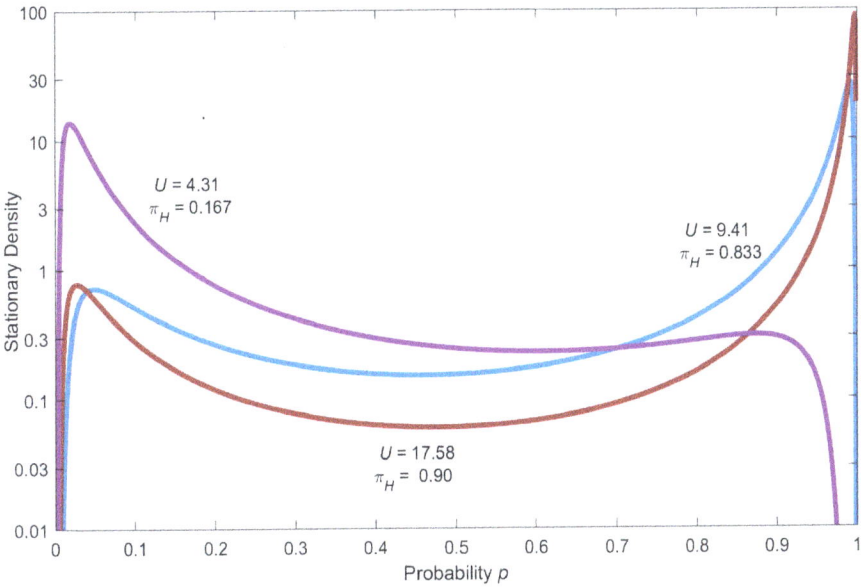

The higher uncertainty associated with rarer states induces higher average volatility. Low-drift bear markets are rarer than high-drift bull markets. Hence, rational learning predicts that bear markets should be more volatile on average than bull markets, as is widely observed. A notable exception is the volatile internet-related bubble of the late 1990s, when investors observed high productivity boosts but weren't sure how long they would last. In that case the bull market was perceived as rarer.

High volatility makes investors more cautious in their bets, which can raise risk premia. However, caution per se should not trigger pessimism or surges in volatility. Rational learning views anxiety and fear more as responses to turmoil than as independent causes. Emotional drivers won favor among economists largely because rational drivers seemed incapable of inducing much volatile excess volatility. The previous examples show that they can. Indeed, learning provides a crisper explanation than behaviorism and meshes better with competitive market pressures. Furthermore, it provides a testable hypothesis. If extended calm makes investors irrationally complacent, price volatility might fall below dividend volatility, whereas rational learning rules this out.

Despite such restrictions, models of rational learning offer immense variety. Thanks to the nonlinear relations between the parameters, small tweaks can leave distinct footprints. The distinctiveness suggests that study of market price history might help to decode the driving beliefs. In finance this is known as technical analysis or charting. Critics include most orthodox theorists, who think that current prices encapsulate all information useful for future prediction. Paul Samuelson's famed EMH-heralding paper (1965) prefaced its only reference to chartists with the adjective "deluded".

Granted, every profession induces some behavioral quirks. Perhaps charting is a stress reducer for traders akin to smoking or stroking worry beads. Still, the scale and intensity amaze. Chemical engineers don't study alchemy. Aeronautical engineers don't study incantations for magic carpets. Why then do financial traders get lured to stuff that most finance professors dismiss?[10] Surely this merits more scholarly attention.

20

Disaster Risks

We saw in the previous chapter that the prospect of rare disasters—low probability π_L of a highly negative drift μ_L—can induce substantial excess volatility and turbulence even when growth is fundamentally sound. This chapter ties two-drift models to empirical evidence, extends the models to let drift vary over a continuum, and incorporates model uncertainty. The adjustments make simulations of market price and volatility look more realistic.

Robert Barro and Jose Ursua (2008) did a banner study of disaster risk, encompassing 7616 observations on real per capita consumption or GDP that I will combine as the findings were close. Defining disaster as a cumulative decline of least 10%,[11] they estimated probabilities $p_{HL} \approx 0.0366$ of switching from normal to disaster and $p_{LH} \approx 0.283$ of switching back. In subsequent work, Barro and Ursua (2012) expanded the data set to 8769 yearly observations. The average durations in the fuller set were 25 years in normal times and 3.64 years in disasters, which imply $p_{HL} \approx 0.040$ and $p_{LH} \approx 0.275$. They estimated an average yearly growth rate of 1.88% with standard deviation 6.24% and mean drawdown during disaster of 21.1%.

Here is a first pass at fitting a two-drift Markov switching model to the data. Solve $(\lambda \pi_L, \lambda \pi_H) \approx (0.04, 0.275)$ to estimate $\lambda \approx 0.315$ and $\pi_H \approx 0.872$. Divide the mean log drawdown of 0.237 by the average duration to estimate $\mu_L \approx -0.0652$. Set $\sigma = 0.0624$ to match the standard deviation.

Then reduce the mean percentage growth rate by $\frac{1}{2}\sigma^2$ to estimate $\langle\mu\rangle \approx 0.0169$, which in turn implies $\mu_H \approx 0.0289$.

Figure 20.1 displays a 100-year simulation of Q and ς using the first-pass parameters, assuming a market of Perfect Learners applying discount rate $r = 0.05$. It resembles Figure 19.5 except that the correlation with price is reversed. This is not surprising given that the first-pass $U \approx 3.6$ and π_L are close to the $U \approx 4.3$ and π_H driving Figure 19.5. In a 100,000-year simulation, Q ranges from 28.0 to 35.7 with mean 34.7. ς ranges from 6.3% to 15.6% with a mean of 8.9% and root mean square—that is, the square root of the average variance—of 9.3%.

Since the empirical volatility estimate is based on yearly data, it includes some variance stemming from changes in drift. Correction shaves 85 bps off the initial estimate of σ. Averaging over a year also muddies distinctions between normal times and disaster. In simulations, average disaster duration using the first-pass estimates was three months longer than intended and the annual p_{HL} was 100 bps too high.

Figure 20.1: Q and ς for 100-year simulation given $r = 0.05$ and first-pass Barro-Ursua fit $(\sigma, \mu_H, \mu_L, \lambda, \pi_H) = (0.0624, 0.0289, -0.0652, 0.315, 0.872)$

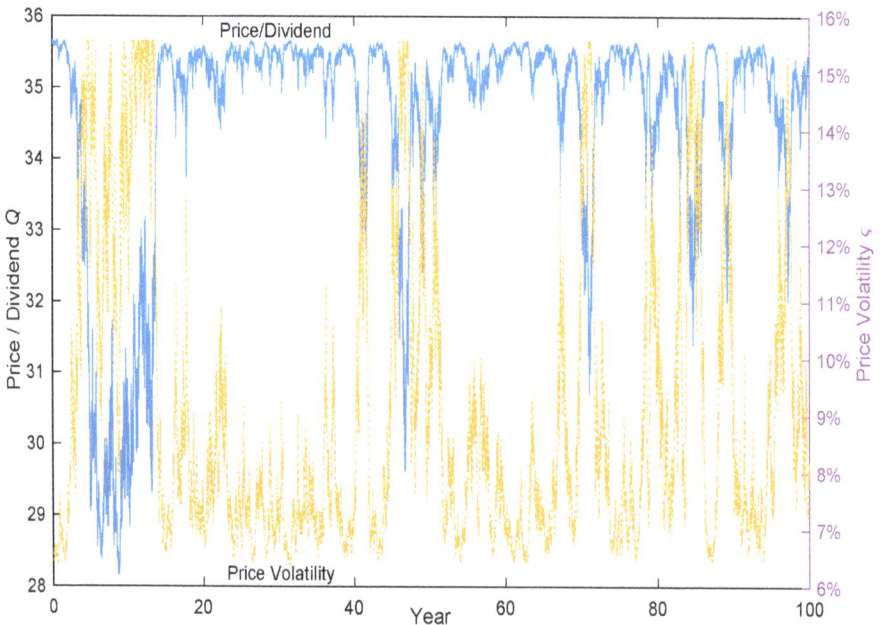

To improve the fit, I conducted a grid search over gradations as fine as 0.001.[12] The best fit is $\mu_L \approx -0.122$, $\lambda \approx 0.251$, and $\pi_L \approx 0.047$, with an induced $\sigma \approx 0.0545$ and $\mu_H \approx 0.0242$. The low-drift downturns are nearly twice as intense as in the first-pass estimates but less than half as common. The average crisis duration of 4.19 years is close to the average duration of the disasters Barro and Ursua measured. However, the average 85 years between crisis regimes is more than triple the average time between measured disasters, which suggests that most of the latter do not signify a truly downward trend.

Figure 20.2 displays a 100-year simulation using the best-fitting parameters. Compared with Figure 20.1, the calm and turbulent periods are far more bifurcated and the surges in volatility are far more intense. In a 100,000-year simulation, Q ranges from 23.1 to 35.2 with mean 34.6. ς ranges from 5.5% to 33.9% with mean 7.3% and root mean square 8.6%. The information potential roughly triples to $U \approx 14.4$.

Figures 20.1 and 20.2 share two features that Figures 19.1-19.5 do not, namely the lower range of Q and the lower ς. Neither feature is crucial to

Figure 20.2: Q and ς for 100-year simulation given $r = 0.05$ and best-fitting parameters $(\sigma, \mu_H, \mu_L, \lambda, \pi_H) = (0.0545, 0.0242, -0.122, 0.251, 0.953)$

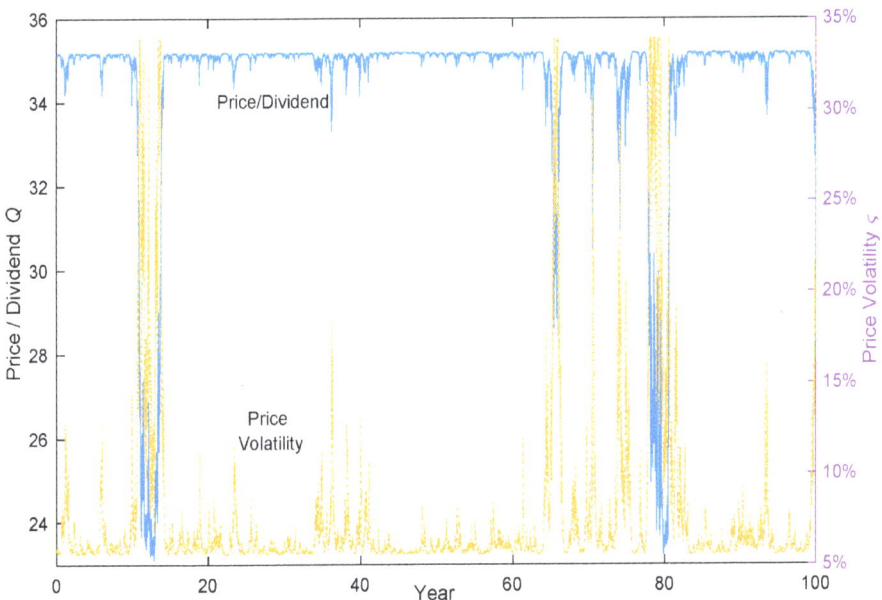

the switching regime. Change in r will shift the range of Q, since it acts in (19.1) like the opposite change in drifts. For example, if we reduce r for Figure 20.2 by 40 bps and rerun the simulations, the minimum, maximum, and mean Q rise by 16% each for a range of 26.7 to 40.9 with a mean of 40.2, while the range and mean of ς scarcely change. As for the minimum ς, we can raise it by allowing for equity leverage. Gross corporate profits are more volatile than revenues or payments for labor and supplies, and net profits after taxes and debt repayments are even more volatile. A simple though crude way to model this scales up all ς by the ratio of corporate dividend volatility to GDP volatility.

The best explanator of the long-term upward trend is almost surely a reduction in $r - \langle \mu \rangle$ through some combination of milder crises and faster mean growth.[13] Barro and Ursua (2008, Table 3) report an average 140 bps faster growth for 1947-2006 compared to 1870-1947. All else being equal, a change of that magnitude would boost Q by about 70%. If r decreased by 40 bps in the same period, Q would more than double.

Although I will not model a secular trend here, I will relax the restriction to two drifts. Imagine that drift equals a constant μ_{MAX} less a random gamma disturbance with shape D and mean μ_{GAM} that resets at rate λ. What parameter combinations can best generate the statistics reported by Barro and Ursua? Recall from Figure 12.1 that $D \ll 1$ makes most gamma outcomes cluster near zero while stretching a dispersed minority well above the mean. The probability of the latter is on the order of D. To approximate the best two-drift fit, σ and λ should lie near their two-drift values, μ_{MAX} should lie near μ_H, μ_{GAM} should lie near $\pi_L (\mu_H - \mu_L)$ and D should lie near π_L.

Another grid search confirms this. In multimillion-year simulations, the combination $\sigma \approx 0.0520$, $\lambda \approx 0.309$, $\mu_{MAX} \approx 0.0287$, $\mu_{GAM} \approx 0.0120$, and $D \approx 0.101$ fits the data best. For this distribution, there is 51% chance of a gamma outcome under 1 bp and 65% chance of an outcome under 10 bps. Of the 17% of outcomes that exceed the mean of 120 bps, the conditional mean is 642 bps and the standard deviation is 699 bps. If the estimates are correct, a severe crisis with mean contraction of 10% or more per year should start on average once in 150 years.

When Brownian drift can take a continuum of feasible values after reset with density π, the Core Learning equation (18.1) indicates that

$$dp(\mu) = p(\mu)\frac{\mu - E}{\sigma}dy + \lambda\big(\pi(\mu) - p(\mu)\big)dt .$$ (20.1)

To calculate the risk-neutral Q_p, let's retrace the method used for two drifts. Given a drift μ of log dividends, an investment of $\overline{Q}_\mu$ returns dividends at rate 1 and grows at expected percentage rate $\mu + \frac{1}{2}\sigma^2$. In an instant dt, there is probability λdt of reset shifting the fair price-to-dividend ratio to Q_π; otherwise the fair ratio stays $\overline{Q}_\mu$. Applying the approximate discount $1 - rdt$ on future returns, equilibrium prices satisfy

$$\overline{Q}_\mu \approx 1 \cdot dt + \big(1 - rdt\big)\big(1 + \mu dt + \tfrac{1}{2}\sigma^2 dt\big)\big((1 - \lambda dt)\overline{Q}_\mu + \lambda dt \cdot Q_\pi\big).$$ (20.2)

First-order conditions require $(\lambda + J(\mu))\overline{Q}_\mu = 1 + \lambda Q_\pi$ for the pure dividend yield $J(\mu) \equiv r - \frac{1}{2}\sigma^2 - \mu$, with solution

$$Q_p = \big\langle \overline{Q} \big\rangle_p = \frac{\langle q \rangle_p}{1 - \lambda\langle q \rangle_\pi} \quad \text{for} \quad q_\mu \equiv \frac{1}{\lambda + J(\mu)} .$$ (20.3)

Equation (19.1) is a special case for two drifts. Again, a negative value implies infinite valuation. To keep valuations finite, all feasible μ must be less than $\lambda + r - \frac{1}{2}\sigma^2$ and $\langle J \rangle_\pi$ must be significantly positive. Our gamma reset parameters meet both constraints.

Figure 20.3 displays a simulation of Q and ς using the best-fitting gamma-reset parameters. The display covers 1000 years because the results are too uneven to identify a single representative century. The maximum and mean Q are close to their counterparts in Figure 20.2, as are the mean and minimum ς. However, the minimum Q is much lower and the maximum ς is much higher. In a 100,000-year simulation, Q ranges from 11.6 to 35.0 with mean 34.1. ς ranges from 5.3% to 319% with mean 7.6% and root mean square 8.9%.

Figure 20.3: Q and ς for 1000-year simulation given $r = 0.05$ and gamma-reset $(\sigma, \mu_{MAX}, \mu_{GAM}, D, \lambda) = (0.052, 0.0287, 0.012, 0.101, 0.309)$

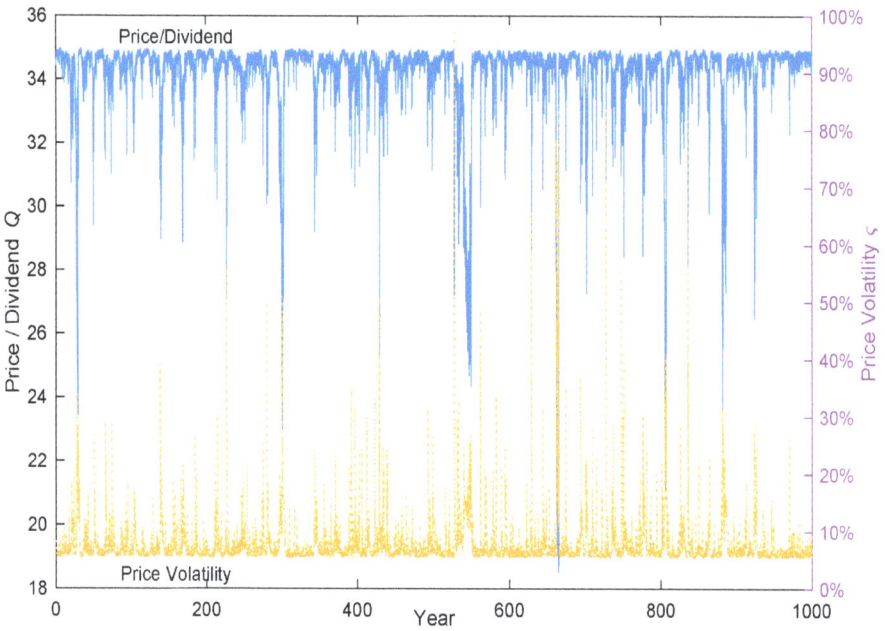

As the gamma extremes occur so rarely, a fairer comparison adjusts for likelihoods. In a 100,000-year simulation, 0.8% of Q for the gamma-reset model fall below the minimum 24.5 for the two-drift model; their mean was 19.6. Similarly, 0.3% of ς for the gamma model exceeded the maximum 38.7% for the two-drift model; their mean was 57.7%.

In all three reset models, price volatilities can be highly volatile in the short-term but mean-revert in the long term. Those are the two most striking features of GARCH behavior. Exponential GARCH (Nelson 1991) fits the volatilities best, since it focuses on percentage changes and allows for extra surges in response to losses. The estimated coefficients are highly significant even in 10-year simulations, However, GARCH treatment of the volatility of volatility is flawed. It presumes that the variance of market prices responds deterministically to past noise, when in fact volatility responds to changing skewness and can be influenced by tiny, hard-to-measure doubts. This favors models with stochastic volatility of volatility.

Model Uncertainty

Barro and Ursua (2012, Figure 2) reported the dispersion of disaster sizes grouped in bands 250 bps wide. Figure 20.4 overlays this with the corresponding averages from million-year simulations. Interestingly, the gamma and two-drift models generate nearly the same distribution. The biggest relative misfit comes at the high end. No empirical drawdown of GDP or consumption exceeded 75%, whereas 2.2% of two-drift disasters and 2.5% of gamma disasters have drawdowns that large. However, the empirical data understates the worst cases for equity holders, e.g., their complete expropriation in Russia during 1913-21, when consumption and GDP fell 62% and 71% respectively.

The most visible misfit with the empirical data is the reversal within the 10%-15% range. Some of this could be sampling error, as the standard deviation in that range is at least 200-300 bps.[14] We could mitigate the discrepancies by incorporating a dispersion target for best fit.

Figure 20.5 displays an analogous chart for the duration of disasters in years. The two reset models generate similar distributions, and again the empirical fit is uneven. Both models predict that slightly more than half of all disasters should last 2-3 years, yet only 38% of recorded disasters did so. They also predict that 2%-3% of disasters should last longer than the recorded maximum of 10 years.

Since gamma reset allow countless drifts, it might be expected to generate substantially more dispersion than the two-drift models. Figure 20.5 shows that it does not. One reason is that yearly averages moderate some shorter-term extremes. The main reason is that dividend noise dominates the differences in underlying drivers.

Suppose that learners misinterpret gamma resets as two-drift resets and vice versa. Equity markets will favor the best short-term predictors, with profits hinging less on objective news than on the reactions of the crowd. If the wrong beliefs currently dominate, they will predict the crowd's reactions better than the right beliefs do. How do we know which sets of beliefs currently dominate the markets? By studying market data. That's where chartists are right and orthodox theorists are wrong.

Figure 20.4: Disaster sizes in Barro and Ursua (2012) and two models

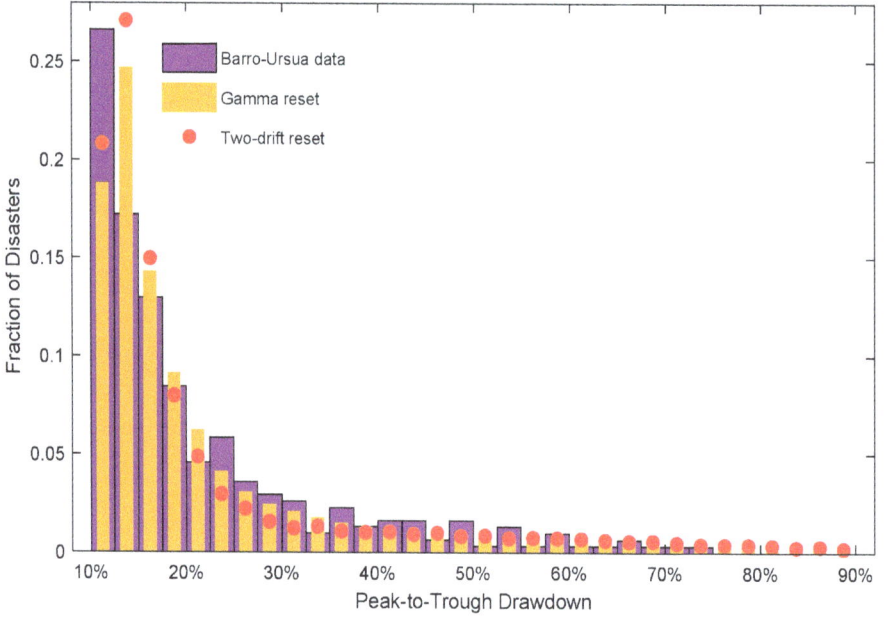

Figure 20.5: Disaster durations in Barro and Ursua (2012) and two models

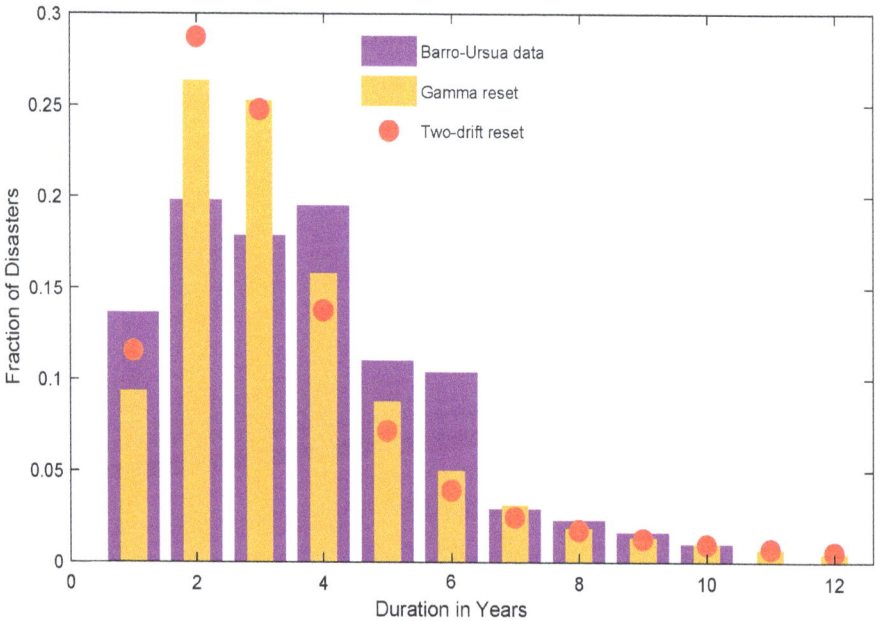

Rival beliefs vying for dominance add volatility, since variance of aggregate beliefs equals mean individual variance plus variance across the individual means. For illustration, suppose an Almost-Perfect Learner (APL) know that a gamma-reset process prevails with $\sigma = 0.052$ and have narrowed down to three possible values $\{b_i - \delta_i, b_i, b_i + \delta_i\}$ for every other parameter. The b_i match the best-fit values reported earlier. The δ_i equal 0.01 for the mean drift, 0.05 for D, 0.005 for μ_{GAM}, and 0.1 for λ. Suppose APL assigns equal initial priors, accords a once-per-century chance of reset from a uniform distribution, and updates beliefs daily.

Figure 20.6 displays Q and ς for a 1000-year simulation. Although the underlying risk drivers are the same as for Figure 20.3, Q fluctuates far more in normal times and normal ς is higher. In a 50,000-year simulation, Q ranged from 10.5 to 56.0 with mean 37.6 and standard deviation 5.7, while ς ranged from 6.1% to 220% with mean 10.1% and standard deviation 4.2%. The model uncertainty makes the price charts look much more realistic, with less regularity from one century to the next.

Figure 20.6: Q and ς for 1000-year simulation given $r = 0.05$, risks similar to Figure 20.3 and market of Almost-Perfect Learners

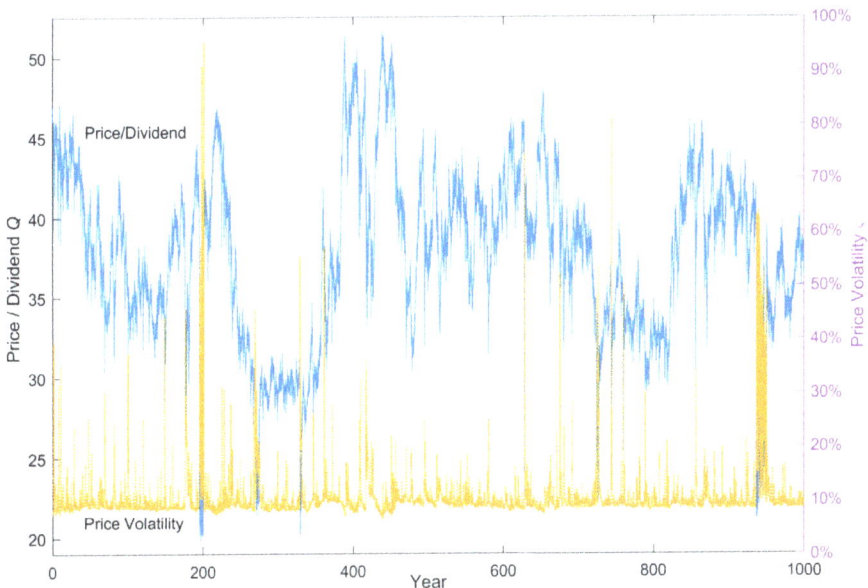

21

Risk Premia

Why is the equity risk premium so high? As noted in Chapter 17, standard models posit a constant relative risk aversion (CRRA) γ high enough to account for it. When all risks are normal, γ must exceed 10. Anticipation of rare disasters justified $\gamma \approx 3$ according to Barro and Ursua (2012), but Nakamura, Steinsson, Barro and Ursua (2013) upped that to $\gamma \approx 6$ after adjusting for post-crash rebounds.

To better comprehend the implications, imagine a lottery ticket that offers a 10% chance of infinite riches. If you would bet everything you own on it, your implied $\gamma \leq 1$. Willingness to pay plummets for higher γ. Are you willing to pay half your wealth? Then your implied $\gamma \approx 1.15$. A quarter of wealth? Then your implied $\gamma \approx 1.37$. As Figure 21.1 shows, willingness to pay shrinks from 10% of wealth when $\gamma = 2$ to 1.1% of wealth when $\gamma = 10$.

In these models, wealth includes the discounted value of lifetime future wages and benefits. Since relatively few people have the means to forfeit 10% of that upfront, the thresholds in Figure 21.1 potentially apply to a majority. However, equity markets are driven by the minority who do have such means. Their wealth is effectively divided into two components: a safety buffer managed with high risk aversion and a much larger discretionary surplus managed with low risk aversion. While the bounds between those components aren't fixed, they change too sluggishly to justify a high average investment γ.

Figure 21.1: Willingness to pay for 10% chance of infinite wealth given relative risk aversion between 2 and 10

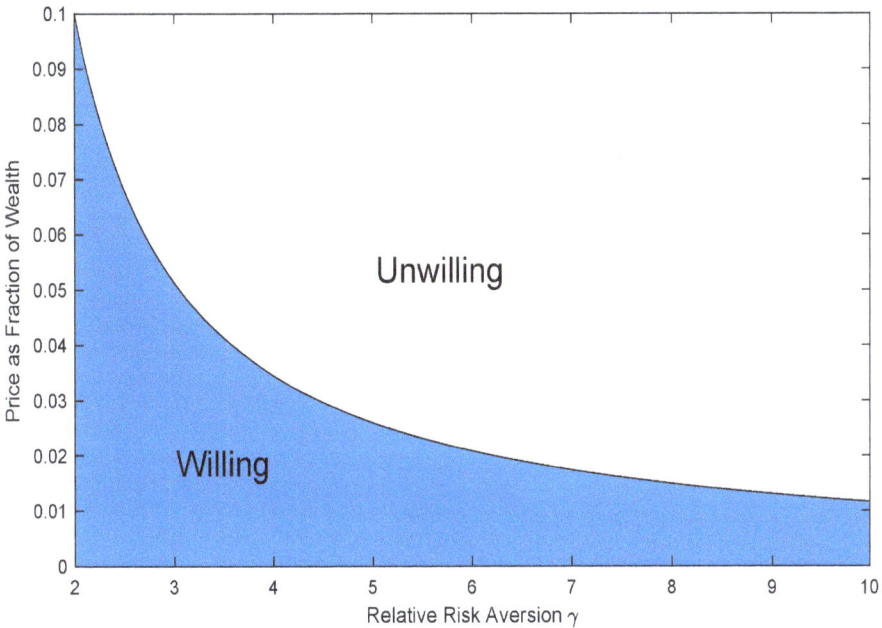

Furthermore, the models described above assume iid risks, where crashes are instantaneous. When we allow for protracted crises and mid-course adjustment, high CRRA is even harder to defend. To see this, let expected utility take the standard time-additive CRRA form

$$\int_0^\infty \frac{\left\langle c(t)^{1-\gamma}\right\rangle - 1}{1-\gamma}\, e^{-\rho t} dt\,,$$

where c denotes consumption. In an instant dt given known drift μ, a starting dividend of 1 is expected to rise to $\left\langle \exp(dx)\right\rangle \approx 1 + \left(\mu + \tfrac{1}{2}\sigma^2\right)dt$ for a marginal expected utility multiplier of $\left\langle \exp((1-\gamma)dx)\right\rangle$.[15] Adjustment modifies the equilibrium condition (20.2) to

$$\bar{Q}_\mu \approx 1\cdot dt + e^{-\rho dt}\left\langle e^{(1-\gamma)dx}\right\rangle\left(e^{-\lambda dt}\bar{Q}_\mu + \lambda dt\cdot Q_\pi\right)$$
$$\approx 1\cdot dt + \left(1 - \rho dt\right)\left(1 + \mu^* dt + \tfrac{1}{2}\sigma^{*2}dt\right)\left[\left(1 - \lambda dt\right)\bar{Q}_\mu + \lambda dt\cdot Q_\pi\right],$$

where $\mu^* \equiv (1-\gamma)\mu$ and $\sigma^* \equiv |1-\gamma|\sigma$. This is equivalent to replacing r, μ and σ in (20.2) and (20.3) with ρ, μ^* and σ^* respectively. The solution is

$$Q_p = \frac{\langle q \rangle_p}{1 - \lambda \langle q \rangle_\pi} \quad \text{for} \quad q_\mu \equiv \frac{1}{\lambda + \rho - \frac{1}{2}(\gamma-1)^2 \sigma^2 + (\gamma-1)\mu}. \quad (21.1)$$

Recall that for $\gamma = 0$, μ was limited in how high it could get without driving valuations infinite. For $\gamma > 1$, μ is limited in how low it can get, as the expected depreciation on the associated security gets too painful to warrant holding it.

This is bad news for fitting the Barro-Ursua data. No reset with gamma shock or any other potentially unbounded shock is feasible. How about a two-drift model? Using the best-fitting parameters with $\rho = 0.01$, there is no finite solution unless $\gamma < 2.61$. Raising ρ relieves the pressure only slightly; even $\rho = 0.04$ requires $\gamma < 3$.

Moreover, all solutions to (21.1) for $\gamma > 1$ make the price-to-dividend ratio fall and the dividend yield rise as growth prospects improve. The inversion makes equity prices less volatile than dividends in normal times and highly negatively correlated with dividends during crises. In short, equities behave more like safe-haven assets than risky plays on growth. The result is effectively baked in by the assumption that equities lay claim on all consumption.

Figure 21.2 displays a 100-year simulation of Q and ς for $\gamma = 2$, $\rho = 0.01$, and the remaining parameters taken from the best-fitting two-drift model. There is extreme bifurcation between calm and turbulent periods. In a 100,000-year simulation, Q ranged from 49.6 to 102 with mean 52.0, while ς ranged from near-zero to 42.6% with a mean of 5.3% and root mean square of 12.5%. Over 80% of Q lay within one point of its minimum, in which case ς averaged 4.0%. The contrasts with real-life

Behavior looks more realistic when equity dividends are modeled as ℓ-times leveraged consumption proxied by a power function ac^ℓ. This replaces $\gamma - 1$ in (21.1) with $\gamma - \ell$. Q flips back into positive correlation with dividend growth provided $\ell > \gamma$. Indeed, we regain the solution (20.3) when $\ell = \gamma + 1$.

Figure 21.2: Q and ς for 100-year simulation given
$$(\rho,\gamma,\sigma,\mu_H,\mu_L,\lambda,\pi_H)=(0.01,2,0.0545,0.0242,-0.122,0.251,0.953)$$

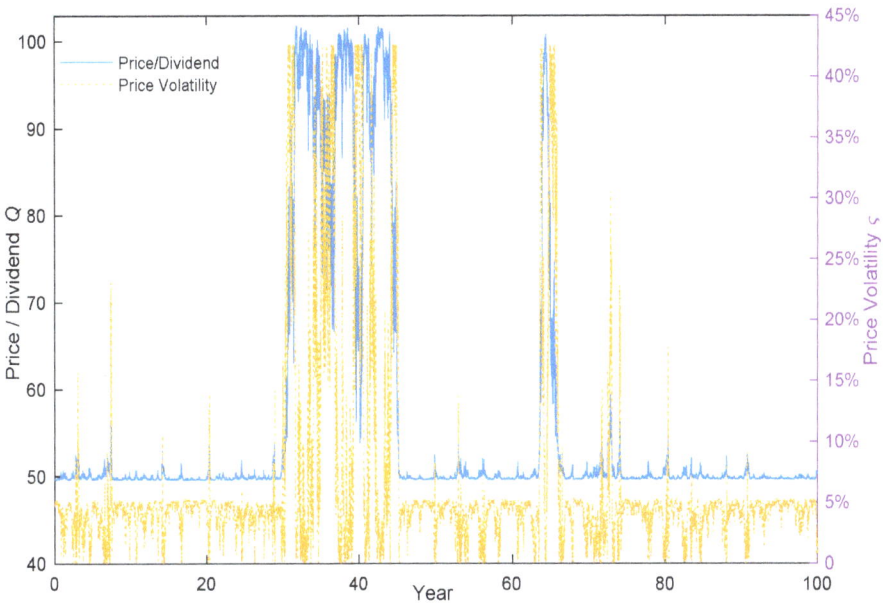

For additional flexibility, time-additive utility can be replaced with recursive utility, which relaxes the intertemporal tradeoffs implied by γ, and investments can be allowed in risk-free bonds. As Jessica Wachter (2013) has shown, risk diffusion and careful choice of parameters can generate far more verisimilitude than the simple equations considered here. For related research, consult her survey with Jerry Tsai (Tsai and Wachter 2015).

While this work is admirably sophisticated, the results seem fragile. Good fits typically set both the intertemporal elasticity and γ implausibly high. The power function approximation to leverage fails badly for high ℓ. The presumed tight relation between equity dividends and aggregate consumption is suspect, as is the market's ability to quickly identify ever-changing trends.

Below I introduce a far easier way to account for a high risk premium. Instead of viewing equity investment as a concerted attempt to smooth aggregate consumption, suppose we treat it as a casino game. The core insight is that the mean belief E must exceed the market price F to persuade rational gamblers to take a net long position in equities.

To analyze this, let's start by comparing an equity's current dividend valued at e^x with a future dividend an instant dt forward. Taking into account both the expected percentage growth rate $\mu_i + \frac{1}{2}\sigma^2$ and the time discount rate ρ, each gambler i values the future dividend as $e^x + e^x(\mu_i + \frac{1}{2}\sigma^2 - \rho)dt$ in current funds. Suppose the comparable market price is $e^x + e^x(F + \frac{1}{2}\sigma^2 - \rho)dt$. If each gambler places the full Kelly bet $k_i K(\mu_i - F)/(\sigma^2 e^x)$ given capital share k_i, the aggregate market bet is $K(E - F)/(\sigma^2 e^x)$ dividends. In equilibrium this should match the number N of dividends, which requires

$$F = E - \frac{Ne^x}{K}\sigma^2 .$$

When beliefs differ, some will want to hold the dividend while others want to short it, but disagreement per se won't boost aggregate demand. The market price F must still drop below E to induce positive holdings. As for the discount, it makes sense for this to rise with dividend variance σ^2 and with the ratio of dividend value Ne^x to investible capital K.

Next let's extend this reasoning to an equity priced at $P = Qe^x$ with volatility ς. Let P_F denote the market-clearing price for the equity an instant dt forward, let P_i denote gambler i's view on the equity's value dt forward, and let $P_E \equiv \sum_i k_i P_i$ denote the market consensus value. Each full Kelly bet aims to hold $k_i K(P_i - P_F)/(P_F^2 \varsigma^2 dt)$ equities. The aggregate bet of $K(P_E - P_F)/(P_F^2 \varsigma^2 dt)$ should match the number of equities in equilibrium, in which case

$$\frac{P_E - P_F}{P_F} = \frac{NP_F}{K}\varsigma^2 dt . \tag{21.2}$$

In words, the risk premium rate over the next instant is $\varsigma^2 NP_F/K$. If we identify NP_F with aggregate value of equities and K with total capital invested in equity markets, then $NP_F \approx K$ and the risk premium is approximately ς^2. This result is attractive in several ways:

- It ties the equity risk premium to the market price volatility ς rather than the lower volatility σ of dividends or consumption.
- Since ς is often on the order of 15%, this can easily account for risk premia on the order of 200 bps even when γ is low.
- Since valuation is negatively correlated with the risk premium, surges in volatility depress the associated Q below the expected value of fully certain $\overline{Q}_\mu$.
- It allows the risk premium to vary with market volatility, which in turn makes prices fluctuate more than they would otherwise and raises the ratio of $\langle \varsigma^2 \rangle$ to σ^2.

Empirical Support and Simulations

Can (21.2) account for observed risk premia? Let's start with a benchmark calculation that harks back to Chapter 17. As a first approximation, the mean dividend rate $Q_\pi^{-1} \approx \rho + \text{RiskPrem} - G$, where RiskPrem denotes the premium over the rate of time preference ρ [16] and G denotes the mean percentage growth rate of dividends. We can estimate RiskPrem as the mean excess return on equities over short-term Treasury bills. According to (21.2), this should approach the average variance $\langle \varsigma^2 \rangle$ of aggregate equity returns.

Figure 21.3 offers empirical support. Drawing on data from Barro and Ursua (2012, Table 2), it compares mean excess returns and the variance of returns for both OECD countries and non-OECD countries from 1870 to 2009 and for the subperiod 1946-2009. For the OECD, the two measures nearly match in both periods. For non-OECD countries, the mean excess returns were roughly half of variance in both periods. One explanation of the lower non-OECD ratio is that their stock markets were less developed and absorbed less of speculative wealth.

One caveat is that the variances in Figure 21.3 are calculated using yearly data, which overstates the variance of pure noise by the variance of the underlying yearly drifts. Hence mean excess returns in the OECD likely exceed $\langle \varsigma^2 \rangle$ in apparent violation of (21.2).

Figure 21.3: Mean excess equity returns over T-bills versus variance of equity returns for OECD and non-OECD countries 1870-2009

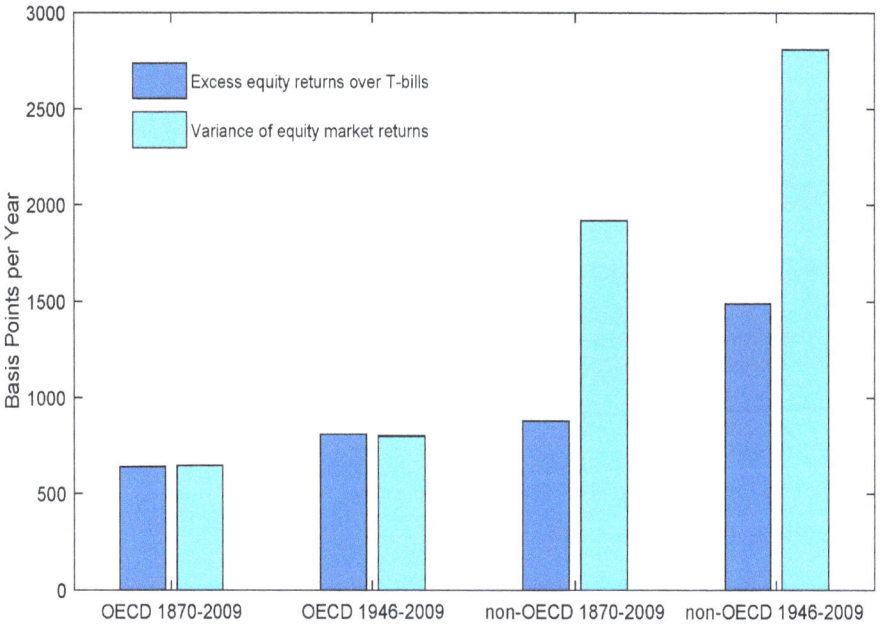

More disturbingly, (21.2) seems to explain at best about half of the historic risk premium in the highly developed capital markets of the US. With long-term Q_π^{-1} on the order of 300 bps, ρ on the order of 100 bps, and G on the order of 200 bps, RiskPrem seems to be on the order of 400 bps. For consistency with (21.2), annual stock market volatility should be on the order of 20%. In fact, the variance of S&P 500 monthly returns has averaged only 200 bps per year. Figure 21.4 charts 3-year and 10-year moving average variance on a log scale, using online data from Shiller (2018). Apart from the 1930s, variance has rarely exceeded 200 bps on a 10-year horizon, while volatility has rarely exceeded 20% on a three-year horizon.

Can the risk premium get so high during occasional S&P 500 disasters that it boosts the average to 400 bps? At first glance the answer is no, because surges in crisis are offset by the retrenchments in normal times. However, these surges and retrenchments add extra volatility to prices, which in turn boosts the average risk premium. While an extra 200 bps seems unlikely, it is hard to gauge the magnitude without simulations.

Figure 21.4: Historical variance and volatility of S&P 500 based on monthly returns

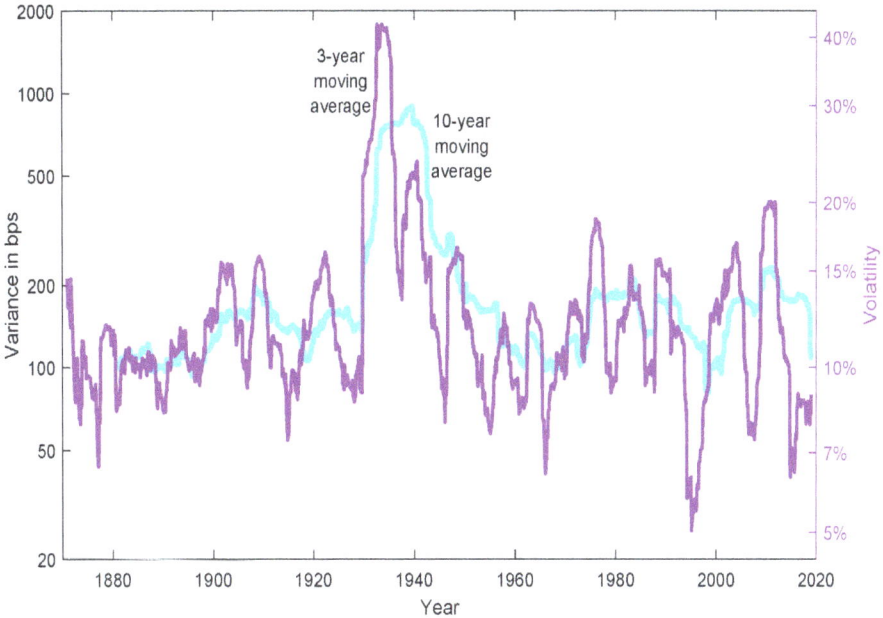

Heterogeneity adds additional complexity. As hard as it is to predict the flux of our own beliefs, it is even harder to predict the flux of others' beliefs and their capital-weighted disagreements with each other. Those beliefs and disagreements have more influence on price volatility than our own. This makes betting on equities riskier than betting on current dividend growth alone.

The extra risks are why (21.2) relates the risk premium to equity price volatility instead of dividend volatility. Still, it might not fully compensate for the perceived risks in gauging others' reactions. Hedging out future risks might not fully compensate either, as no gamblers understand exactly what risks they're taking to what degree. Between transaction costs, deference to others and fear of the unknown, equity traders will tend to bet less than the full Kelly associated with perfect learning. If fractional Kelly trims all bets, the equilibrium risk premium is modified to

$$\frac{P_E - P_F}{P_F} = \frac{NP_F}{fK}\varsigma^2 dt \ . \tag{21.3}$$

In well-developed equity markets with $NP_F \approx K$, the instantaneous risk premium over ρ should be approximately ς^2/f. If f drops too low, bolder investors or momentum traders will step in and drive its effective value higher. However, half Kelly might be sustainable since it is nearly as efficient as full Kelly. Half Kelly implies a risk premium of $2\varsigma^2$, which is easy to reconcile with historical averages.

For a fuller treatment, let's modify our pricing models under reset to allow an endogenous risk premium and fractional Kelly gambling. On the learning side, fractional Kelly slows credibility updates (20.1) to

$$dp(\mu) = p(\mu) f \frac{\mu - E}{\sigma} dy + \lambda\left(\pi(\mu) - p(\mu)\right)dt . \qquad (21.4)$$

On the pricing side, replacing a constant r with the risk-adjusted $\rho + \varsigma_p^2/f$ changes the equilibrium price relations (20.2) to

$$Q_p \approx 1 \cdot dt + \left(1 - \left(\rho + \varsigma_p^2/f\right)dt\right)\left(1 + \left(E + \tfrac{1}{2}\sigma^2\right)dt\right)\left\langle \hat{Q}_{p+dp}\right\rangle, \qquad (21.5)$$

which implies

$$\left(\rho + \varsigma_p^2/f - E - \tfrac{1}{2}\sigma^2\right)Q_p = 1 + \left\langle dQ_p\right\rangle/dt . \qquad (21.6)$$

This can be simplified further by using Itô's Lemma and (21.4) to evaluate $\left\langle dQ_p\right\rangle$ and combining with (19.4.) However, even with a simple two-drift model, the resulting second-order differential equation is highly nonlinear and not readily solvable. I found it easier to discretize p over a fine grid and express the vector $\left\langle dQ_p\right\rangle$ as a matrix product ΠQ. Given a reasonable candidate vector ς, (21.6) is readily solved for Q. When ς is seeded with σ and Q is updated by a quarter of the recommended changes at each step, the system typically converges to solution within 100 iterations. Occasionally a starting Q blows up and needs to be reseeded.

A bigger problem is that Q_p is highly sensitive to its local steepness and convexity; a coarse grid cannot do the calculations justice. The following results use a grid of 5000 points, with five times tighter spacing above $p = 0.98$ than below. Why very high p merits finer attention will become clear in the next chapter. To compute Π, I applied (21.4) to every starting p on the grid for 200,000 daily Brownian dy and mapped the resulting $p + dp$ to their nearest neighbors on the grid.

We can then iterate on (21.4) and (21.5) until $Q_p \approx \hat{Q}_p$ for all p.[17] As a benchmark, I will use the μ_H, μ_L, λ and π_H estimates that best fit the Barro-Ursua data along with $\rho = 0.01$. I also set $f = 0.67$, whose CRRA analogue is $\gamma \approx 1.5$. My only other change bumps σ to 0.1 in order to align it more with dividend volatility than with consumption volatility.

Figure 21.5 displays a 100-year simulation of Q and ς. Its closest fixed-r analogue is the best-fitting two-drift model of Figure 20.2, which shows similar peak volatilities. However, Figure 21.5 looks far more market-like. Its Q spans a wider range with less bifurcation between calm and turbulent periods. In a 100,000-year simulation, Q ranged from 19.6 to 43.1 with mean 40.1. Price volatility ς ranged from 10.3% to 28.6% with a mean of 15.7% and root mean square of 16.4%. While each parameter in this simulation can be challenged, every plausible modification generates simulations that look far more realistic than Figure 21.2.

Figure 21.5: Q and ς for 100-year simulation given
$(\rho, \sigma, f, \mu_H, \mu_L, \lambda, \pi_H) = (0.01, 0.1, 0.67, 0.0242, -0.122, 0.251, 0.953)$

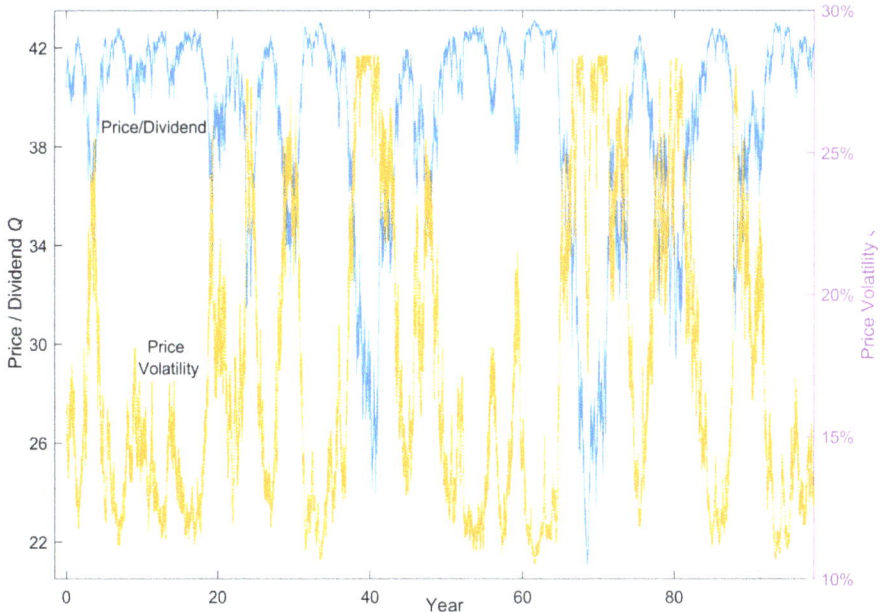

22

Pricing Challenges

To see how fractional Kelly generates high price volatility with high discount rates, let's look at the underpinnings of Figure 21.5. All its novelties stem from adjustments in Q for market risk; beliefs p about dividend growth are unaffected. Figure 22.1 charts the adjusted Q as a function of p. Not only are the endpoints Q_0 and Q_1 very low but also the curve is highly convex. In contrast, a constant discount rate would leave Q_p linear in p. The inward bow reflects the dependence of excess volatility on $p(1-p)$. The uncertainty depresses Q_p well below the linear combination of Q_0 and Q_1 that would apply under risk-neutrality.

At first glance, convexity affects only the intermediate values of the curve and not the endpoints. In fact, it lowers the whole curve because the valuations are interdependent. Anticipating migration, both Q_0 and Q_1 have to discount the convexity-reduced values of intermediate Q_p, and when Q_0 and Q_1 drop the intermediate values must drop too. A similar argument applies to every intermediate segment; e.g., convexity between $p = 0.5$ and $p = 0.6$ pulls down $Q_{0.5}$ and $Q_{0.6}$.

In decision theory, a depressed response to uncertainty is known as uncertainty aversion and treated as an add-on to risk aversion. Since standard expectations are linear in subjective probabilities, uncertainty aversion is often confined to realms of hazy forecasting and unquantifiable "Knightian uncertainty". In contrast, our approach explains the aversion in

Figure 22.1: Q versus p for Figure 21.5

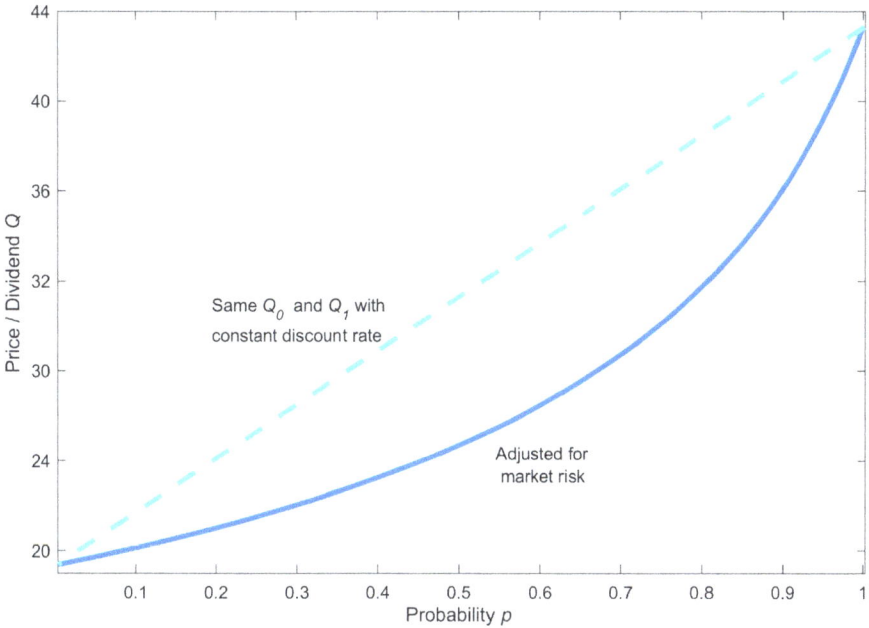

Same Q_0 and Q_1 with constant discount rate

Adjusted for market risk

a wholly rational, quantifiable way. Gamblers aren't marking down $Q_{0.5}$ because they inherently dislike high uncertainty. They're marking it down because the uncertainty makes the asset more volatile than its near-term stream of dividends.

It is hard to quantify uncertainty aversion since we rarely observe p directly. For example, suppose we reinterpret the prices in Figure 21.5 as fair values given a constant discount rate. One way to generate the required Q_0 and Q_1 is to set $\lambda = 0.097$ and $r = 0.0497$ while retaining all the other best-fit parameters. Another way keeps λ unchanged but sets $\mu_L \approx -0.307$ and $r = 0.0458$. Using either method we would infer that the market not only imposes a significant risk premium but also exaggerates the expected drawdown in crisis.

The feature that most clearly distinguishes Figure 21.5 from its fixed-r analogues is its more volatile volatility outside crisis That is implicit in Figure 22.1. As p declines from 1 to 0.95, Q drops by over 10%, over 700 bps more than if the discount rate were constant. Only a steep rise in volatility can account for that. The risk premium ς^2/f at $p=0.95$ is 500 bps, 350 bps higher than its minimum 150 bps under certainty. It continues rising down to $p=0.7$, peaking at 1220 bps, and then gradually declines.

Figure 22.2 charts the risk premium versus p and highlights the extra impact of volatile Q. However, the average height of 700 bps grossly exaggerates the true mean, since most p cluster near 1. In long simulations, 90% of p exceed 0.90, 78% exceed 0.95, and 46% exceed 0.98. Figure 22.3 displays a histogram of the risk premia.

The sensitivity to small $1-p$ doubts is not apparent when charting ς versus Q. That graph looks nearly parabolic with a slight tilt to the left and we can easily find a constant-r specification that mimics it.[18] As in Figure 22.1, a rational Kelly gambler could easily be interpreted as applying a high fixed discount rate with an exaggerated view of downside risks. Alternatively, the same gambler could be interpreted as moved by "animal spirits" that frequently change the effective discount rate and induce needlessly excess volatility. This helps us appreciate how orthodox and behaviorist schools of finance can draw strikingly different conclusions.

Figure 22.2: Risk premium ς^2/f versus p for Figure 21.5

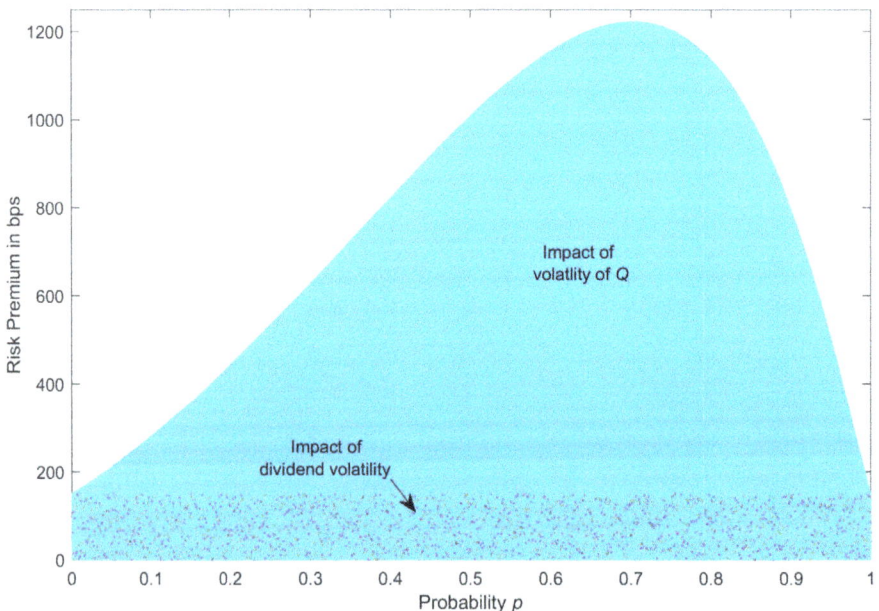

Figure 22.3: Histogram of risk premia for Figure 21.5

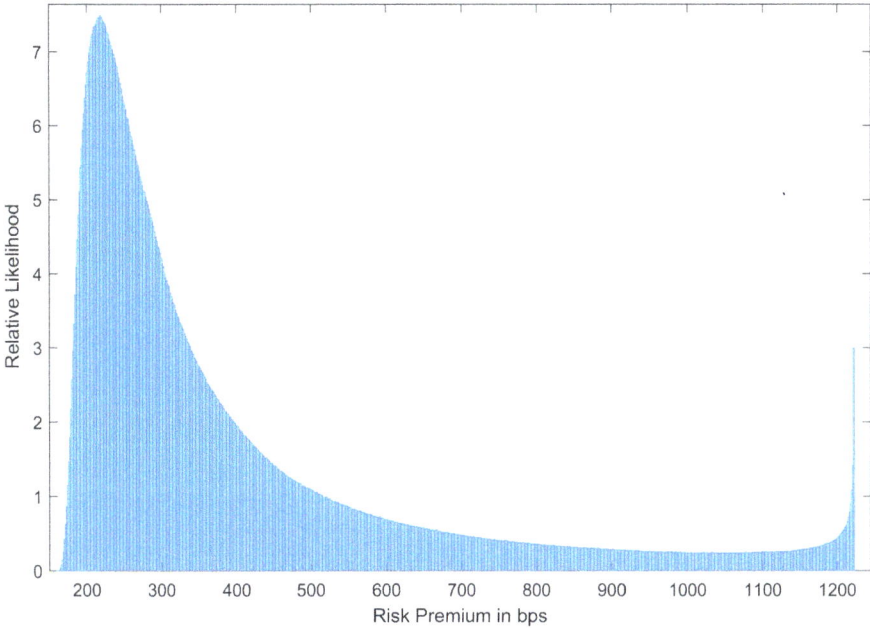

One shortcoming of this model is the restricted range of Q and ς. No matter how bad the news, Q never falls below 45% of its highs and ς never reaches 30%. The main cause is the presumption that the disaster drift is fixed. What happens when we substitute a gamma-reset model?

Computation gets far more difficult when a fixed shock is replaced by $N \gg 1$ possible shocks, as our beliefs will span a N-dimensional simplex. The curse of dimensionality forces us to coarsen the search, which aggravates concerns about rounding errors and convergence. To simplify, I restricted Q_p to a function $Q_{E,V}$ of E and V, as the first two cumulants capture nearly all the short-term variation. Indeed, E alone provides a good approximation, since V is highly negatively correlated with it, except at very low E where V stabilizes or declines. In benchmark simulations of p, the top 99% of E were -0.92 correlated with the associated V.

To estimate the transition matrix Π for $Q_{E,V}$, I simulated 400,000 years of daily dividends 250 trading days per year. Every simulated day I tracked a 500-point p and recorded its E and V, for 100 million pairs in all. I separated the pairs into 2000 bins ordered by magnitude of E. These bins were equal in size except for the lowest 1% of E , where bins contained only a fifth as many elements in order to trim their range and distinguish better between degrees of disaster. I then subdivided each bin into five equal-sized smaller bins ordered by V. I computed Π as the average 5-day transition rates across the 10,000 bins and adjusted the dividend and discount rates accordingly.

Like in the two-drift model, I seeded ς with σ and an initially constant Q and then iterated between updates of Q and recalculations of ς . Unlike in the two-drift model, my attempts to solve (21.6) directly did not induce convergence, perhaps due to noise in Π . However, iterations via (21.5) converged slowly, and the results were relatively robust to variations in the numbers of bins.

My first set of simulations with the best-fit gamma reset parameters retained $\rho = 0.01$, $\sigma = 0.1$, and $f = 0.67$. The solution drove the maximum Q under 20 and the average risk premium over 600 bps. I then redid the simulations with full Kelly $f = 1$. Figure 22.4 displays a 100-year simulation.

Figure 22.4: Q and ς for 100-year simulation given $(\rho, \sigma, f) = (0.01, 0.1, 1)$ and gamma-reset $(\mu_{MAX}, \mu_{GAM}, D, \lambda) = (0.0287, 0.012, 0.101, 0.309)$

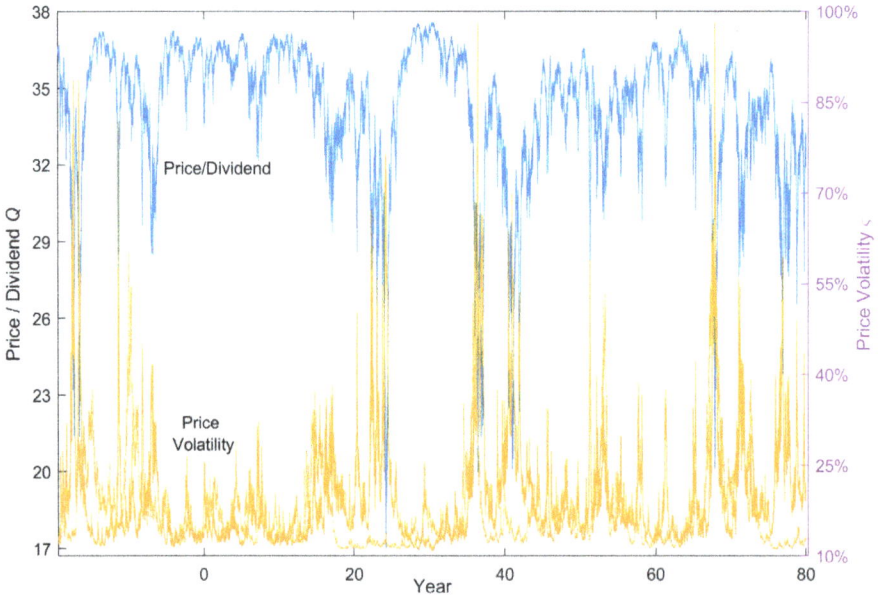

This brought typical price behavior back toward the levels in Figure 21.5. The mean Q was 34.2 with a high of 37.6. The implied average risk premium was 417 bps, again reflecting a steep increase on small fears of disaster. The most striking differences from Figure 21.5 involve peak contractions and volatilities. With gamma reset and $f = 1$, Q was 2% likely to fall below fell below 20 and 1% likely to fall below 15. ς was 5% likely to exceed 42%, 2% likely to exceed 60%, and 1% likely to exceed 75%.

These responses seem too strong to describe real-life markets. One possibility is that the Barro-Ursúa focus on drawdowns overstates the risks. Nakamura et al (2013) found that rebounds after disaster recover roughly half of the drawdowns. One crude way to incorporate that is to double the gamma shape D. This roughly doubles the mean Q in simulations and trims the average risk premium by about 150 bps.

Why does mild anticipation of rare disasters depress Q so much even when risk aversion is low? The core driver is the uncertainty it stokes. Perhaps the downturn will continue, in which case the perceived odds of uncertainty will soar. Perhaps the downturn will reverse, reassuring that it was just an unlucky streak in an otherwise healthy trend. Since market price will likely respond very strongly to the noise in subsequent news, rational gamblers demand an extra premium for holding on.

Momentum Trading

The equilibrium computations are challenging. I cannot solve a general model without iterating hundreds of times. To the extent that real-life markets make similar iterations, it might be more lucrative to bet on short-term trends than to bet on their long-term destination.

Heterogeneity makes equilibrium even harder to compute. The market risk premium responds not just to the volatility expected when everyone thinks alike but also to the volatility induced by disagreement. Current volatility in turn depends partly on projections of future volatility. Perfect equity pricing would require the anticipation of everyone's contingent doubts and disagreements along with the contingent capital weights to place on them.

Responding to this impossibility, we resemble the proverbial drunk who searches for keys where the light is best. Since the best light we have on volatility and pricing comes from the market's own history, we are bound to project past trends into the future. Mindful of the risks, we may trim our bets and slow adjustment. However, slow adjustment lures momentum traders to identify the current trend and bet that it continues. As noted in Chapter 1, Ilmanen (2011) found that momentum trading worked better than any other simple investment strategy.

To behaviorists, widespread momentum trading demonstrates that mindless herding prevails. From a learning perspective, momentum trading is a step in the rational evolution of capital markets. Its very success tends to exhaust the opportunities it is exploiting. Consider Commodity Trading Advisors (CTAs), which focus on algorithmic pursuit of price trends that last a few weeks or months. A recent article describes their fall from grace:

> "Robotic traders now manage about $1 out of every $3 held in the world's $3 trillion hedge fund industry, including models that use inputs like company's profitability, trends in volatility or shifts in economic cycles to make trading decisions. But trend followers keep it simple, identifying when to enter and exit trades by back-testing price trends against decades of data. Problem is, they aren't very good at responding to surprises… Ultimately, the speed of markets these days can easily confound the historical price trends at the heart of the approach…
>
> The turn against CTAs comes a decade after they rose to fame for rewarding investors with average returns of 21 percent in 2008…. Institutional investors piled in, sending assets soaring 69 percent to $270 billion in the decade to 2017, according to data provider Eurekahedge. The strategy hasn't really delivered. Between 2008 and 2018, Société Générale's main trend-following index made only 3.7 percent, compared with an average gain of 62 percent for hedge funds and a more than tripling in the S&P 500, including dividends." (Kumar 2019)

If markets are speeding up with automation, then effective deference $1-f$ is falling, which implies that average Q should rise. That's consistent with what we observe. However, correlation is not causation, and even if a rise was mostly justified it might have overshot its support. Momentum trading always brooks a tendency to overshoot.

Even when markets bet close to full Kelly, they encourage myopia since long term trends are hard to predict and corrective news arrives slowly. Excess price volatility exacerbates this. To see why, let's compare the expected values of short-term and long-term insight for two magical machines:

- Overnight Oracle identifies daily one of 100 iid components q_i of tomorrow's change ΔP in market price.
- Future Prophet identifies a halving or doubling of true long-term value that is expected to correct within T years.

For Overnight Oracle, the optimal Kelly bet per unit wealth generates an expected profit of $\langle q_i^2 \rangle / \mathrm{var}(\Delta P - q_i) \approx 0.01$ per trading day. For Future Prophet, assuming constant log price convergence at yearly rate $\log(2)/T$ and yearly price volatility ς, the optimal Kelly bet per unit wealth generates an expected profit of approximately $(\log(2)/\varsigma T)^2$ per trading year. When $\varsigma = 15\%$, T must be less than three years for Future Prophet to offer more value, whereas if ς were halved the maximum T favoring Future Prophet would double to six years.

Granted, an efficient portfolio strategy could use both machines. Future Prophet requires little daily maintenance and imposes few transactions costs. Bear in mind too that major mispricing tends to correct unevenly, with most of the catchup occurring in a relatively short interval. A gambler informed by Future Prophet might do best by waiting for evidence of a surge before investing heavily. That's a form of momentum trading, even though Future Prophet is purely value-oriented.

Asset allocation tends to boost momentum. Since many portfolio managers exaggerate their forecast expertise, potential backers judge in part by past results. As a one-time wonder, Future Prophet might not garner any investment until the price had largely converged as predicted.

Incentives for financial news and analysis also favor a short-term focus. Each morsel of current news has an immediacy that gives a potential trading edge to those who receive and process it quickest. Distinguishing correct interpretations from misinterpretations takes longer. So does infusing new perspectives from research and refining them through reasoned debate. While this likely adds more long-term value than a flash news report, it is harder for publishers to capitalize on. The most notable exceptions involve publications by the research departments of brokers or fund managers, which aim to advertise their superior know-how to clients or induce rivals to buy what they already hold.

Immediacy explains why professional sports have suffered far less than professional music from the pirating of recordings. Once the final score is known, full game replays lose most of their excitement and are savored only by coaches and scouts. In contrast, most listeners savor new music better on rehearing.

For equities, every trading moment provides the latest score. Imagine a security paying dividends tied to the difference between observed global temperatures and historic benchmarks. Its core value would hinge on future global warming. Yet most of the related financial news would focus today's price action and unusual current weather.

In short, momentum trading complements fundamental valuation. It mitigates risk aversion and speeds adjustment to new trends. However, it may also distract from socially more important endeavors and seems prone to overshoot. What new risks do rational gamblers create when they track each other's tracking and bet on their projections into the future? How might policymakers mitigate these risks without blunting the wisdom of market crowds? These questions seem likely to challenge the theory and practice of finance for decades to come.

Notes

Part I: Bridging Divides

1. Photo US Navy, retrieved from cdn.wallpapersafari.com/w/ZKhy4A.

2. en.wikipedia.org/wiki/Quechuan_languages. Quechuan grammar also distinguishes between observations that are first-hand, inferred or reported by someone else. For simplicity, this book ignores such distinctions. It treats all public information as shared without distortion. Even with this restriction, rational learning often generates sharp disagreements. Allowing for doubts about the reliability of evidence would compound this.

3. This sketch of physical turbulence ignores the vortices that grow, spin off and dissipate. This book focuses on rational learning about scalar random variables, where it is impossible to generate vortices. I do not know whether learning about multi-dimensional variables generates something akin to vortices.

4. Darwin knew that many traits are not blended, with male/female distinctions the most obvious. He conjectured that pangenesis passed on some traits without variation. To reconcile pangenesis with variety, Darwin inferred that stress on male and female reproductive systems induced numerous mutations. The inference was logically consistent but seemed hard to reconcile with Mendelian genetics.

5. Asexual bacteria have evolved ways to maintain genetic variation, including ingestion of genes from other bacteria (Pierce 2012, Chapter 20). However, sexual reproduction maintains variation more reliably.

Part II: Harold's Casino

1. In a few percent of simulations of Harold's Surprise, likelihoods evolve a small mound between 0.6 and 0.9. The associated transitions start quicker than the norm but finish slower. Why? Because they make it easier to believe that θ has jumped at least 0.1 and harder to believe that θ is close to 1. Even in these cases the dominant feature is bifurcation between bigger mounds near 0.5 and 1. This makes E respond far more intensely to news than if beliefs clustered near 0.75.

2. For better consistency between this price-taking behavior and profit maximization, we can reinterpret any large capital K_i as subdivided between multiple bots who share i's beliefs. Agents who know they can influence market price will shade their bids, which to some extent works like the risk aversion and social deference discussed in Chapter 10.

3. I chose 15 TDs for Figure 9.1 to facilitate identification of the median and choose 16 TDs here to facilitate identification of capital shares. Otherwise the shift is unimportant.

4. Surely the brain is far more decentralized than standard neural networks suggest. There is no evidence of a central depot for major nerve endings, much less a tiny mind-within-the-mind homunculus that supervises the depot and directs all traffic. Centralization is also prone to overheating and blowouts. Since an inter-threaded market greatly econo-mizes on information flows without inducing permanent monopolization by big winners, I suspect that the analogy can provide useful insight for neuroscience. This is not to claim that brains exactly mirror inter-threaded markets.

5. Futures are often far easier to trade than the corresponding present claims thanks to simpler documentation and cheaper transfers. The advantages are huge in commodities markets, where settlement of the present claim often requires delivery of the commodity or provision for storage. Where settlement costs are significant, our analysis is better reframed as a comparison of long-term futures to short-term futures.

Part III: Credit Markets

1. Volume-weighted rates come from Giesecke et al (2010, Table 2 and Figure 1) for 1866-1899, Hickman (1953, Tables A-2 and A-17) for 1900-1943, Atkinson (1967, Table 21) for 1944-1965, Moody's (Ou, Hamilton and Cantor 2003) for 1970-1993 and Moody's (Ou et al 2018, Exhibit 39) for 1994-2017. I equated the missing volume-weighted rates for 1966-69 to the corresponding tiny (0-0.12%) issuer-weighted rates. Issuer-weighted rates come from Moody's (Ou et al 2018, Exhibit 2) for 1920-1980 and an average of Moody's and S&P (Vazza and Kraemer 2018, Table 1) for 1981-2017.

2. When average default risk is gamma-distributed, defaults have a negative binomial distribution with scale proportional to the pool size. Assuming pool size is constant, the best-fitting shape is approximately 0.6 on the full sample and 0.8 on the post-1900 subset. Adjusting for S&P's reported pool sizes since 1920 (the value weightings used before 1920 make pool size misleading), the best fit is 0.8 on the full sample and 1.0 on the post-1900 subset.

3. For proof that this generates the target distribution, see Osband (2011, 216-218). The only other diffusions generating this distribution have volatility sv^τ for constant τ. Empirical fits to default history favor $\tau \approx 0$.

4. More precisely, $s \approx 1.2$ is the best fit for either metric when D_0 is trimmed to 0.6 as explained below. At $D_0 \approx 0.8$, the best fit for s stays 1.2 for the yearly-change metric but drops to 1.0 for the cycle-length metric. My fitness measure was the average RMSE in at least 500 simulations.

5. With N iid issuers, the variance of default rates equals $1/N$ of the variance for a single issuer. Average variance is higher and the pool effectively smaller when N fluctuates or value-weighting raises concentration.

6. The best fit for the 35.5 observed cycles is $D_0 \approx 0.6$ and $s \approx 1.6$, while the best fit for the 0.7 median absolute change in log default rates is $D_0 \approx 0.6$ and $s \approx 1.9$. In simulations, the value $s \approx 1.7$ minimizes the average sum of relative errors, as the cycle-based estimator is more precise than the volatility-based estimator for the sovereign sample.

7. A neater formulation preserves (12.1) but focuses on beliefs about the aggregate default rate Nv. It replaces v with Nv, E with NE, and V with N^2V. The replacement also allows us to mix credits of different risk, provided we treat high-risk credits as bundles of independent low-risk components that default if even a single component fails.

8. Let diag(v) denote the diagonal matrix with diagonal elements v_j. and let π denote the matrix of diffusion rates π_{ij}, where $\pi_{ii} \equiv -\sum_{j \neq i} \pi_{ij}$. Folding in reset at vector rate Λ from a probability vector P_0, the vector S of survival rates satisfies $dS/dt = (\pi - \mathrm{diag}(\Lambda + v) + \Lambda P_0')S$.

9. To convert the diffusion (12.3) into a daily transition matrix Π, I set a daily $dt = \frac{1}{250}$ that allows 250 trading days per year, halved it M times until all transitions were confined to immediate neighbors and then squared the intraday matrix M times to restore a daily interval.

10. S&P reports only aggregate CCC-C default rates, while Moody's separates Caa from Ca-C but doesn't break it down by tier prior to 1998. In post-1998 data, default rates for Caa2 averaged twice the Caa1 rates and Caa3 rates averaged 1.6 times Caa2 rates. I imputed missing Moody's data by assuming these ratios held throughout and by assigning the headline rate to Caa2. I imputed the S&P counterparts similarly with an additional assumption that the CCC rate is 80% of the CCC-C rate.

11. I mapped ω to grades assuming geometric spacing between the thresholds implicit in Figure 14.1. The lower bound for triple-A is too close to zero to matter. To reduce the simpact of very weak credits, I treated all C-grades as triple-C. Without that compression, the reported RMSEs would be higher. Despite a conservative mapping, the average ω is 1.38, which boosts the average default rate $\langle \omega v \rangle$ in the sample to 248 bps. I will explain the differential later.

12. Correlations between single-A and single-B or lower grades are less than 0.17. When default rates in one grade are regressed against default rates in other grades, only adjoining grades are highly significant.

13. Mapping sovereign ω using the same procedures as for corporate ω, I estimated an average ω of 1.17, which boosts the average default rate $\langle \omega v \rangle$ in the sample to 316 bps. Again, I will explain the differential later.

14. In matrix form, $u \approx 2(I + P_{trunc})^{-1}(p_{default} + p_{withdraw})$, where I is an identity matrix and P_{trunc} is a truncated P without the columns $p_{default}$ and $p_{withdraw}$ for default and withdrawal.

15. The faster reversion is consistent with S&P setting a higher ω threshold between single-B and triple-C grades than Moody's does.

16. Suppose a credit in grade i has expected ratings life L_i. If it transits to grade j in one year, its expected rating life becomes $L_j + 1$. Weighting each transition by probability P_{ij} and aggregating, $L = \mathbf{1} + LP_{trunc}$ for vector L of L_i and unit vector $\mathbf{1}$, which implies $L = (I - P_{trunc})^{-1}\mathbf{1}$.

17. Given the huge uncertainties, I have experimented with simpler models of ratings migration. The best fit for mixed diffusion/reset sets $\langle\omega\rangle = 2$, $D_0 = 0.55$, $s = 0.25$, and $\Lambda = 0.01$. Unfortunately, it systematically understates the likelihood of two-grade moves and greatly overstates reversion rates from C grades. Empirical transition rates tend to decline log-linearly with migration distance, whereas diffusions with reset imply log-quadratic decay with floors. The most plausible explanation is that credits of the same grade can systematically differ in their stability.

18. Formally, Basel II associated credit ratings with multipliers that converted asset values to risk-adjusted values and imposed an 8% capital requirement on the latter (Basel Committee 2004, 2005). Table 15.1 merges these two steps.

19. Assuming log-linearity any two estimates—e.g., the 0.48 log span of tiers and 0.29 lower bond on investment grade—suffice to set the rest.

20. The conversion can't be perfect since the empirical Π is not fully positive definite. I obtained an excellent approximation by (i) taking the matrix square root, setting negative values to zero, and rescaling the rows to sum to one, (ii) repeating this procedure five more times to obtain the 64th root Π_{64}, and (iii) estimating the weekly transition matrix assuming 50 trading weeks per year as $\Pi_{50} = I + \frac{64}{50}(\Pi_{64} - I)$. The absolute differences between $(\Pi_{50})^{50}$ and Π averaged 0.2 bps and never exceeded 4 bps.

21. When assets are uncorrelated, this measures diversity as the reciprocal of the relative portfolio weights squared. Substituting market shares for portfolio weights yields the Herfindahl-Hirschman Index (Herfindahl 1950, Hirschman 1964) used to measure market competition.

Substituting relative abundances for portfolio weights yields the Simpson index (1949) of biological diversity.

22. My simulations of $v^* = \omega v$ let ω and v evolve independently. This roughly halves the 0.55-0.6 shape of the ω and v distributions alone. Since the history used to model aggregate sovereign v is likely influenced by fluctuations in ω, my models arguably understate the shapes of both v and v^*. However, the understatement in simulations is slight—less than 0.1—and does not alter the main conclusions.

23. For a stylized example, suppose default risk is a power function of the debt-to-GDP ratio b and that b grows at the rate of the fair credit spread. Osband (2011, 49-51) shows that default is inevitable within a fixed time period and yet interest rates can stay low until shortly before the end.

Part IV: Equity Markets

1. Specifically, $\xi \equiv (1 - 1/\psi)/(1 - \gamma)$ where ψ is the intertemporal elasticity of substitution given recursive Epstein-Zin (1989) preferences. When utility is time-additive, $\psi = 1/\gamma$ and $\xi = 1$. Empirical research leans toward $\psi < 1$, with most estimates between 0 and 0.5 but some as high as 2. This leaves wide scope for estimates of ξ although most favor $\xi > 0.2$ given moderate γ.

2. Each distribution is discretized over a grid of 100 possible g values spaced evenly between 0 and 400 bps.

3. The kurtosis of a t-distribution with υ degrees of freedom is $6/(\upsilon - 4)$. A t-distribution with high υ is nearly normal. The best-fitting t-distribution for log dividends sets $\upsilon \approx 6$.

4. See Liptser and Shirayev (1977) for a rigorous treatment that remains the most comprehensive to date. However, their treatment does not reveal the cumulant hierarchy implied by rational learning or the turbulence associated with it.

5. A variance-gamma distribution can be generated by either mixing normal random variables with gamma-distributed variances or by taking the difference of two gamma-distributed random variables having the

same shape. It has a taller head and fatter tails than a normal distribution and can be skewed.

6. In general, $\sqrt{V_0} \ll \Delta\mu \ll \sigma$ induces a kind of slow-motion turbulence where $\Delta\mu$ is a huge shock to beliefs but takes a long time to detect.

7. Given normality $\log\langle e^{dx}\rangle = K(1) = \langle dx\rangle + \frac{1}{2}\langle(dx)^2\rangle = \kappa_1 dt + \frac{1}{2}\kappa_2 dt$.

8. Applying (19.4), mean price volatility is 13.7%, 60 bps less than for Example 3. It is 104 bps higher than for Example 3 (17.3% versus 16.2%) when computed in the more traditional way as the square root of the average variance.

9. Given Brownian motion with drift $\kappa_1(p)$ and unit variance $\kappa_2(p)$, stationarity requires $h^*(p) \propto \kappa_2^{-1}(p) \cdot \exp(2\int \kappa_1(p)\kappa_2^{-1}(p)dp)$. Applying this to (19.2) yields (19.5).

10. Andrew Lo and Jasmina Hasanhodzic (2009, 2010) are welcome exceptions.

11. Despite this definition, Barro and Ursua recorded one 9.5% decline (for the US in 1914) as a disaster. Setting a 9.5% threshold would slightly intensify the rational turbulence simulated in this chapter.

12. For every tested parameter combination, I ran at least one Monte Carlo simulation of 100,000 years or longer with weekly updates. For every candidate (μ_L, λ, π_L), I adjusted σ and μ_H to generate the average yearly percentage growth and standard deviation reported by Barro and Ursua. I then compared the average disaster intensity and two switching probabilities to the Barro-Ursua values and sought the combination that minimized the sum of squared relative errors. Starting with a coarse grid, I gradually narrowed the range while making the grid finer within that range. To mitigate the impact of noise, my closing rounds computed the best fits in million-year simulations, repeated the process a dozen times and selected the modes.

13. A blend of dividends, GDP growth and other data might provide better guidance on profitability than dividends alone. To the extent that markets track non-dividend data, the reported Q will span a broader range than the model Q. However, this can't come close to accounting for the huge range of historical price-to-dividend ratios displayed in Figure 17.1.

14. Given N independent disasters with probability p of falling in a given range, the standard deviation of the estimator is . While the data cover 308 disasters, most of the reported consumption disasters overlap with reported GDP disasters, and many disasters across countries are highly correlated due to world war or global depression.

15. More precisely, consumption $\exp(dx)$ over the next instant is valued at the ratio $\exp(-\gamma dx)$ of marginal utilities, with expectation $\langle \exp((1-\gamma)dx) \rangle$.

16. In standard theory, the equity risk premium includes an additional component, namely the excess of r over a risk-free bond rate. I have not attempted to analyze this.

17. For more robustness to rounding errors, I updated (21.6) assuming weekly (5 trading day) holding periods, with replaced the daily Π with Π^5 and effectively multiplied the other daily parameters by 5.

18. With a fixed discount rate, excess volatility in (19.4) is a quadratic expression in p divided by a gradually increasing Q_p. The quadratic is parabolic while the divisor tilts it left.

References

Atkinson, Thomas R. 1967. *Trends in Corporate Credit Quality*. National Bureau of Economic Research, Studies in Corporate Bond Financing. New York: Columbia University Press.

Allen, Robert C. 2009. *The British Industrial Revolution in Global Perspective*. New York: Cambridge University Press.

Asness, Cliff. 2016. Fama on Momentum. *AQR Insights*, February. www.aqr.com/Insights/Perspectives/Fama-on-Momentum.

Bachelier, Louis. 1900. Théorie de la speculation. *Annales Scientifiques de l'École Normale Supérieure*, 3 (17), 21–86.

Barro, Robert J. 2006. Rare Disasters and Asset Markets in the Twentieth Century. *Quarterly Journal of Economics* 121(3), 823–866.

Barro, Robert J. and José F. Ursúa. 2008. Macroeconomic Crises since 1870. *Brookings Papers on Economic Activity* 39(1), 255–350.

Barro, Robert J. and José F. Ursúa. 2012. Rare Macroeconomic Disasters. *Annual Review of Economics* 4(1), 83–109.

Basel Committee on Banking Supervision. 2004. *International Convergence of Capital Measurement and Capital Standards. A Revised Framework.* Basel: Bank for International Settlements.

Basel Committee on Banking Supervision. 2005. *An Explanatory Note on the Basel II IRB Risk Weight Functions.* Basel: Bank for International Settlements.

Bayes, Thomas. 1763. An Essay Towards Solving a Problem in the Doctrine of Chances. Published posthumously in *Philosophical Transactions of the Royal Society of London* 53, 370–418 and 54, 296–325.

Bernoulli, Daniel. 1738. Specimen theoriae novae de mensura sortis. *Commentarii Academiae Scientiarum Imperialis Petropolitanae 5,* 175–192. Translated by Louise Sommer as Exposition of a New Theory on the Measurement of Risk. *Econometrica* 22(1), 22–36.

Brunt, Liam. 2006. Rediscovering Risk: Country Banks as Venture Capital Firms in the First Industrial Revolution. *Journal of Economic History* 66, 74–102.

Chin, William and Marc Ingenoso. 2007. Risk Formulae for Proportional Betting. In Stewart. N. Ethier and William R. Eadington, eds., *Optimal Play: Mathematical Studies of Games and Gambling.* Reno, NV: Institute for the Study of Gambling, University of Nevada, 541–550.

David, Alexander. 1997. Fluctuating Confidence in Stock Markets: Implications for Returns and Volatility. *Journal of Financial and Quantitative Analysis* 32(4), 427–462.

Dawkins, Richard. 1986. *The Blind Watchmaker.* London: Longman.

de Finetti, Bruno. 1937. La prévision: Ses lois logiques, ses sources subjectives. *Annales Institute Henri Poincaré* 7, 1–68.

Ecke, Robert. 2005. The Turbulence Problem: An Experimentalist's Perspective. *Los Alamos Science* 29, 124–141.

Edwards, Anthony W.F. 2011. Mathematizing Darwin. *Behavioral Ecology and Sociobiology* 65(3), 421–430.

Engle, Robert F. 1982. Autoregressive Conditional Heteroskedasticity with Estimates of Variance of UK Inflation. *Econometrica* 50(4), 987–1008.

Epstein, Larry G. and Stanley E. Zin. 1989. Substitution, Risk Aversion, and the Temporal Behavior of Consumption and Asset Returns: A Theoretical Framework. *Econometrica* 57(4), 937–969.

Fama, Eugene. 1965. The Behavior of Stock Market Prices. *Journal of Business* 38(1), 34–105.

Feynman, Richard, 1965. *Lectures on Physics,* Volume 1. Reading, MA: Addison-Wesley.

Fisher, Ronald A. 1922. On the Mathematical Foundations of Theoretical Statistics. *Philosophical Transactions of the Royal Society of London, Series A* 222, 309–368.

Fisher, Ronald A. 1930. *The Genetical Theory of Natural Selection.* Oxford, UK: Clarendon.

Giesecke, Kay, Francis A. Longstaff, Stephen Schaefer and Ilya Strebulaev. 2011. Corporate Bond Default Risk: A 150-Year Perspective. *Journal of Financial Economics* 102(2), 233–250.

Hansen, Lars P. 2007. Beliefs, Doubts and Learning: Valuing Macroeconomic Risk. *American Economic Review* 97(2), 1–30.

Hansen, Lars P. 2010. Calibration, Empirical Evidence, and Stochastic Equilibrium Models. *Presentation for Institute for New Economic Thinking,* April.

Herfindahl, Orris C. 1950. *Concentration in the Steel Industry.* Dissertation: Columbia University.

Hickman, W. Braddock. 1953. *The Volume of Corporate Bond Financing Since 1900.* National Bureau of Economic Research, Studies in Corporate Bond Financing. Princeton, NJ: Princeton University Press.

Hirschman, Albert O. 1964. The Paternity of an Index. *American Economic Review* 54(5), 761-762.

Hoffman, Philip T., Gilles Postel-Vinay, and Jean-Laurent Rosenthal. 2007. *Surviving Large Losses: Financial Crises, the Middle Class and the Development of Capital Markets*. Cambridge, MA: Harvard University Press.

Ilmanen, Antti. 2011. *Expected Returns: An Investor's Guide to Harvesting Market Rewards*. Chichester, UK: John Wiley.

Kalman, Rudolf E. 1960. A New Approach to Linear Filtering and Prediction Problems. *Journal of Basic Engineering* 82(1), 35–45.

Kelly, John L. 1956. A New Interpretation of Information Rate. *Bell System Technical Journal* 35, 917–926.

Keynes, John M. 1921. *A Treatise on Probability Theory*. London: MacMillan.

Keynes, John M. 1936. *The General Theory of Employment, Interest and Money*. London: MacMillan.

Keynes, John M. 1937. The General Theory of Employment. *Quarterly Journal of Economics* 51(2), 209–223.

King, Mervyn. 2016. *The End of Alchemy: Money, Banking, and the Future of the Global Economy*. New York: W.W. Norton.

Knight, Frank H. 1921. *Risk, Uncertainty and Profit.* Boston, MA: Hart, Schaffner & Marx.

Kumar, Nishant. 2019. One of Wall Street's Most Popular Trading Strategies Is Failing. Bloomberg, February. www.bloomberg.com/news/articles/2019-03-01/one-of-wall-street-s-most-popular-trading-strategies.

Kurz, Mordecai. 1994a. On Rational Belief Equilibria. *Economic Theory* 4, 859–876.

Kurz, Mordecai. 1994b. On the Structure and Diversity of Rational Beliefs. *Economic Theory* 4, 877–900.

Kurz, Mordecai. 1997. Asset Prices with Rational Beliefs. In M. Kurz, ed., *Endogenous Economic Fluctuations: Studies in the Theory of Rational Belief*. Berlin: Springer-Verlag, 211–250.

Kurz, Mordecai. 2009. Rational Diverse Beliefs and Economic Volatility. Chapter 8 in Thorsten Hens and Klaus Shenk-Hoppé, eds., *Handbook on Financial Markets: Dynamics and Evolution*. Amsterdam: North Holland, 439–506.

Lamb, Horace. 1932. Address to the British Association for the Advancement of Science. Cited in Michael Tabor, *Chaos and Integrability in Nonlinear Dynamics: An Introduction*. 1989. New York: Wiley, 187.

Liptser, Robert S. and Albert N. Shirayev. 1977. *Statistics of Random Processes*, Volumes 1 and 2. New York: Springer-Verlag. Translation of *Statistika Sluchainykh Protsessov*. 1974. Moscow: Nauka.

Liu, Yang, Quiyang Li, Thorsten Nestmann, and Elena H. Duggar. 2018. *Sovereign Default and Recovery Rates, 1983-2017.* New York: Moody's Investor Service, Data Report, July.

Lo, Andrew and Jasmina Hasanhodzic. 2009. *The Heretics of Finance: Conversations with Leading Practitioners of Technical Analysis.* New York: Bloomberg.

Lo, Andrew and Jasmina Hasanhodzic. 2010. *The Evolution of Technical Analysis: Financial Prediction from Babylonian Tablets to Bloomberg Terminals.* New York: Bloomberg.

McComb, William D. 1990. *The Physics of Fluid Turbulence.* Oxford, UK: Clarendon.

Marcinkiewicz, Józef. 1938. Sur une proprieté de la loi de Gauss. *Mathematische Zeitschrift* 44, 612–618.

Mandelbrot, Benoit B. 1963. The Variation of Certain Speculative Prices. *Journal of Business* 36(4), 394–419.

Mandelbrot, Benoit B. 1997. *Fractals and Scaling in Finance: Discontinuity, Concentration, Risk.* New York: Springer.

Markowitz, Harry M. 1952. Portfolio Selection. *Journal of Finance* 7(1), 77-91.

Martin, Ian W. R. 2013. Consumption-Based Asset Pricing with Higher Cumulants. *Review of Economic Studies* 80(2), 745–773.

Mehra, Rahnish, ed. 2008. *Handbook of the Equity Risk Premium.* Amsterdam: Elsevier.

Mehra, Rahnish and Edward C. Prescott. 1985. The Equity Premium: A Puzzle. *Journal of Monetary Economics* 15(2), 145–161.

Mokyr, Joel. 2009. *The Enlightened Economy: The Economic History of Britain 1700-1850.* New Haven, CT: Yale University Press.

Nakamura, Emi, Jón Steinsson, Robert Barro, and José Ursúa. 2013. Crises and Recoveries in an Empirical Model of Consumption Disasters. *American Economic Journal: Macroeconomics* 5(3), 35–74.

Navier, Claude-Louis. 1822. Mémoire sur les lois du mouvement des fluides. *Mémoires de l'Académie des Sciences de l'Institut de France* 6, 389–440.

Neher, Richard A. and Boris I. Shraiman. 2011. Statistical Genetics and Evolution of Quantitative Traits. *Reviews of Modern Physics* 83(4), 1283–1300.

Nelson, Daniel B. 1991. Conditional Heteroskedasticity in Asset Returns: A New Approach. *Econometrica* 59(2), 347-370.

Osband, Kent. 2001. *Iceberg Risk: An Adventure in Portfolio Theory.* New York: Texere.

Osband, Kent. 2005. Blackjack in the Dark. *Wilmott* 1, 38-41.

Osband, Kent. 2011. *Pandora's Risk: Uncertainty at the Core of Finance.* New York: Columbia University Press.

Ou, Sharon, David T. Hamilton, and Richard Cantor. 2003. *Moody's Dollar Volume-Weighted Default Rates.* New York: Moody's Investor Service, Special Comment, March.

Ou, Sharon, Sumair Irfan, Yang Liu, Joyce Jiang and Kumar Kanthan. 2018. *Annual Default Study: Corporate Default and Recovery Rates, 1920-2017.* Moody's Investor Service, Data Report, February.

Pastor, Lubos and Pietro Veronesi. 2009. Learning in Financial Markets. *Annual Review of Financial Economics* 1, 361–381.

Pierce, Benjamin A. 2012. *Genetics: A Conceptual Approach.* New York: W.H. Freeman.

Poundstone, William. 2005. *Fortune's Formula: The Untold Story of the Scientific Betting System that Beat the Casinos and Wall Street.* New York: Hill and Wang.

Popper, Ben. 2017. Poker-Playing AI is Getting Smarter and the Humans are Getting Tired. *The Verge*, January. www.theverge.com/2017/1/25/14358246/ai-poker-tournament-cmu-libratus-vs-human-losing.

Reinhart, Carmen M. and Kenneth S. Rogoff. 2009. *This Time Is Different: Eight Centuries of Financial Folly.* Princeton, NJ: Princeton University Press.

Reynolds, Osborne. (1883). An Experimental Investigation of the Circumstances Which Determine Whether the Motion of Water Shall Be Direct or Sinuous, and of the Law of Resistance in Parallel Channels. *Philosophical Transactions of the Royal Society of London* 174, 935–982.

Rietz, Thomas A. 1988. The Equity Risk Premium: A Solution. *Journal of Monetary Economics* 22(1), 117–131.

Samuelson, Paul A. 1965. Proof that Properly Anticipated Prices Fluctuate Randomly. *Industrial Management Review* 6(2), 41–49.

Sargent, Thomas J. 2008. Evolution and Intelligent Design. *American Economic Review* 98(1), 5–37.

Savage, Leonard J. 1954. *The Foundations of Statistics.* New York: Wiley.

Savage, Leonard J. 1971, Elicitation of Personal Probabilities and Expectations, *Journal of the American Statistical Association*, 66(336), 783–810.

Sharpe, William F. 1966. Mutual Fund Performance. *Journal of Business* 39(1), 119-138.

Shiller, Robert J. 1989. *Market Volatility*. Cambridge, MA: MIT.

Shiller, Robert J. 2000. *Irrational Exuberance.* New York: Doubleday.

Shiller, Robert J. 2018. U.S. Stock Markets 1871-Present and CAPE Ratio. Retrieved from www.econ.yale.edu/~shiller/data.htm.

Simpson, Edward H. 1949. Measurement of Diversity. Nature 163, 688.

Stiglitz, Joseph E. 2010. An Agenda for Reforming Economic Theory. *Presentation for the Institute for New Economic Thinking*, April.

Stokes, George G. 1851. On the Effect of Internal Friction of Fluids on the Motion of Pendulums. *Transactions of the Cambridge Philosophical Society* 9, 8–106.

Taleb, Nassim N. 2001. *Fooled by Randomness: The Hidden Role of Chance in Life and in the Markets*. New York: Random House.

Taleb, Nassim N. 2007. *The Black Swan: The Impact of the Highly Improbable*. New York: Random House.

Taleb, Nassim N. 2012. *Antifragile: Things That Gain from Disorder*. New York: Random House.

Tetlock, Philip E. and Dan Gardner. 2015. *Superforecasting: The Art and Science of Prediction*. New York: Crown.

Thorp, Edward O. 1960. Fortune's Formula: The Game of Blackjack. *Notices of the American Mathematical Society* 7(7), 935–936.

Thorp, Edward O. 2000. The Kelly Criterion in Blackjack, Sports Betting, and the Stock Market. In Olaf Vancura, Judy A. Cornelius, and William R. Eadington, eds*., Finding the Edge: Mathematical and Quantitative Analysis of Gambling*. Reno, NV: Institute for the Study of Gambling and Commercial Gaming, University of Nevada, 163–214.

Tsai, Jerry and Jessica A. Wachter. 2015. Disaster Risk and Its Implications for Asset Pricing. *Annual Review of Financial Economics* 7(1), 219–252.

U.S. Bureau of Economic Analysis, 2018. Corporate Profits After Tax with Inventory Valuation Adjustment and Capital Consumption Adjustment, and Gross Domestic Product. Retrieved from fred.stlouisfed.org /series/CPATAX and /GDP.

Varian, Hal R. 1985. Divergence of Opinion in Complete Markets: A Note. *Journal of Finance* 40(1), 309–317.

Vazza, Diane and Nick W. Kraemer. 2018. *2017 Annual Global Corporate Default Study and Rating Transitions.* New York: Standard & Poor's, Global Credit Portal, RatingsDirect, April.

Wachter, Jessica A. 2013. Can Time-Varying Risk of Rare Disasters Explain Aggregate Stock Market Volatility? *Journal of Finance* 68(3), 987–1035.

Witte, Lawrence R. and Jason M. Ontko. 2018. *Default, Transition, and Recovery: 2017 Annual Sovereign Default Study and Rating Transitions*. New York: Standard & Poor's, Global Credit Portal, RatingsDirect, May.

Wonham, Walter M. 1964. Some Applications of Stochastic Differential Equations to Optimal Nonlinear Filtering. *SIAM Journal on Control* 2(3), 347–369.

Wright, Sewall. 1932. The Roles of Mutation, Inbreeding, Crossbreeding and Selection in Evolution. *Proceedings of the 6th International Congress in Genetics* 1, 356–366.

Index

www.ingramcontent.com/pod-product-compliance
Lightning Source LLC
Chambersburg PA
CBHW041914190326
41458CB00023B/6258